Birkhäuser

Frontiers in the History of Science

Frontiers in the History of Science is designed for publications of up-to-date research results encompassing all areas of history of science, primarily with a focus on the history of mathematics, physics, and their applications. Graduates and post-graduates as well as scientists will benefit from the selected and thoroughly peer-reviewed publications at the research frontiers of history of sciences and at interdisciplinary "frontiers": history of science crossing into neighboring fields such as history of epistemology, history of art, or history of culture. The series is curated by the Series Editor with the support of an international group of Associate Editors.

More information about this series at http://www.springer.com/series/15796

Davide Crippa

The Impossibility of Squaring the Circle in the 17th Century

A Debate Among Gregory, Huygens and Leibniz

Davide Crippa
Université Paris Diderot
SPHère
Paris, France

Davide Crippa's research for this book has been supported by the postgraduate project n. L30009160: "Bolzano and the foundations of analysis: a sociological exploration," hosted by the Institute of Philosophy of the Czech Academy of Sciences.

Frontiers in the History of Science
ISBN 978-3-030-01637-1 ISBN 978-3-030-01638-8 (eBook)
https://doi.org/10.1007/978-3-030-01638-8

Library of Congress Control Number: 2018962678

Mathematics Subject Classification (2010): 01Axx

Cover illustration: From Oliver Byrne: The First Six Books of The Elements of Euclid

This book is published under the imprint Birkhäuser, www.birkhauser-science.com by the registered company Springer Nature Switzerland AG
The registered company address is: Gewerbestrasse 11, 6330 Cham, Switzerland

Preface

This book is the first outcome of a research programme in the study of impossibility results in mathematics. I began this programme in my doctoral dissertation, of which this book is an abridged and improved version. I am indebted to the many people whose assistance allowed this book to come into existence: first and foremost, Marco Panza, whose helpful and careful guidance made this work possible, and Vincenzo de Risi, who gave me the concrete opportunity to publish it. I also heartily thank Andy Arana, Andrey Bovykin, Abel Lassalle Casanave, Jesper Lützen, Bodo von Pape, Jean-Jacques Szczeciniarz, and David Rabouin for their invaluable suggestions, help and criticism.

I wrote this book as a postdoctoral fellow at the Centre for Science, Technology and Society Studies of the Czech Academy of Sciences. I warmly thank Jan Balon for the opportunity to spend two fruitful years there. I also thank Ladislav Kvasz, Ansten Klev, and Jan Marsalek.

This book was made possible thanks to the financial support of the Centre for Science, Technology and Society Studies of the Czech Academy of Sciences, the ANR-DFG project Formalism, Formalisation, Intuition and Understanding in Mathematics (FFIUM), and the GDR 3398 Histoire des mathématiques.

Finally, I thank Alexander Reynolds and Samuel Eklund for their careful linguistic revisions, and Sarah Goob for her assistance with this publication.

Unless otherwise indicated, all the images in this book are created with the freeware GeoGebra.

Paris, France
August 2018

Davide Crippa

Contents

Introduction 1

1.1 The Quadrature of the Circle and Its Impossibility

In this book, I will study several attempts to prove the impossibility of solving three fundamental problems in geometry by algebraic means: the squaring of the circle, the ellipse and the hyperbola within the mathematical context of Seventeenth Century. All of these problems involve measuring areas or, in modern parlance, evaluating certain integrals. The term "quadrature" reveals the geometrical tradition in which these problems were originally conceived. Within the tradition of Greek mathematics, and in Seventeenth Century geometry as well, to find the area of a figure meant to construct, by geometrical means (the ruler and the compass, in the easiest instances, or by higher curves), a square equivalent to it: "squaring" or "quadrature" are thus just synonyms for designating this geometrical operation. Since the circle, the hyperbola and the ellipse (but not the parabola) are conic sections that possess geometrical centres, I shall refer to them as "central conic sections," and I use the synthetic expression "quadrature of the central conic sections" for the problem of determining their areas.

The story I am going to tell regards a corpus of a few published and unpublished texts, all dating between 1667 and 1676. The *terminus a quo* coincides with the publication of James Gregory's *Vera Circuli et Hyperbolae Quadratura*. This treatise, as the title says, was devoted to the quadrature of the central conic sections, and it sparked a lively interest soon after its publication. The main reason for such acclaim was one of its central results, stating the impossibility of squaring the central conic sections. In modern parlance, Gregory's result states that the general circle, ellipse and hyperbola-measuring integrals have no closed-form, algebraic solutions. The debates after the publication of Gregory's book are interesting not only from a historical point of view, since they constitute the first recorded discussion on an impossibility result in mathematics and contributed significantly to shaping the early history of calculus, but especially from an epistemological one. In

D. Crippa, *The Impossibility of Squaring the Circle in the 17th Century*,
Frontiers in the History of Science, https://doi.org/10.1007/978-3-030-01638-8_1

fact, the question about the solvability of the circle-squaring problem had an explicit epistemological dimension for Gregory, since it was related to the conceptual issue of fixing the boundaries of geometrical knowledge, which, at the time, was represented by Cartesian geometry. If the circle-squaring problem was unsolvable through finite algebraic methods, then the very meaning of exactness in geometry would have to be rethought and its bounds renegotiated.

Gregory's work on quadratures also exerted a great impact on Leibniz's early mathematical thought, especially because of its emphasis on impossibility results. Leibniz was, perhaps in a more pronounced way than Gregory, worried by foundational and philosophical concerns about the limits of Cartesian geometry. For this reason, he was also particularly responsive to the epistemological follow-ups of unsolvability questions in geometry. A tell-tale sign of Gregory's influence on Leibniz's investigation is the last proposition of his *De quadratura arithmetica circuli ellipseos et hyperbolae cujus corollarium est trigonometria sine tabulis* (1676), in which we can find a new, improved proof of the algebraic unsolvability of the quadrature of the central conic sections.

Gregory and Leibniz are thus going to be the main characters of my book, and their theorems about the algebraic impossibility of squaring the central conic sections shall be discussed at length in Chaps. 2 and 3, respectively. Even if the topic has already been explored in the secondary literature, a better understanding of the role exerted by Gregory's impossibility theorem on Leibniz has been made possible only recently, thanks to the publication of the latter's mathematical manuscripts from 1673 to 1676, and particularly the manuscripts devoted to the quadrature of the circle.[1] In this book, I shall also offer a comprehensive study of such rich and novel manuscript material, which forms part of the context of the debate about the impossibility of quadratures during the second half of the Seventeenth Century.

1.2 The Famous Problems of Classical Geometry and Their Impossibilities

The squaring of the central conic sections generalizes the older problem of the quadrature of the circle. This, together with the trisection of an arbitrary angle, the problem of constructing regular polygons and the duplication of the cube (or, more generally, the insertion of two mean proportionals between given segments), was one of the famous problems of ancient Greek geometry.[2]

[1] See Leibniz (1923, VII, 6) and Crippa (2017) for an overview.

[2] This designation became stable probably starting from the Sixteenth or early Seventeenth Centuries, as it appears from the titles of several treatises that mention the three problems together. See, for example, Johannes Molther's *Problema Deliacum de cubi duplicatione* (1619), the subtitles of which runs: "ubi historia problematis praemittitur, et simul nonnulla de Anguli trisectione, heptagoni fabrica, circulique quadratura," or Christiaan Huygens' *De circuli magnitudine inventa* (1654),

Thanks to the circulation of the works of Archimedes and Pappus, in the Sixteenth and Seventeenth Centuries, these age-old problems were hailed as difficult, intriguing and open questions that led many on a wild goose chase in the hope of gaining everlasting glory.[3]

Early-modern mathematicians were, of course, aware that neither of these problems were absolutely unsolvable. As a matter of fact, cube duplication, trisection of the angle and quadrature of the circle had produced several solutions, most of which stemmed from a tradition of research that already had its roots in pre-Euclidean Greek geometry.[4]

What is common to all of these solutions is their non-elementary character, in the sense that they all appeal to assumptions and methods that are not postulated or implicitly used in Euclid's geometry. This comes, however, as a logical necessity: to duplicate a cube, to trisect an angle and to square a circle are all impossible tasks within Euclid's *Elements*. The difficulties associated with the classical construction problems thus depended on the reliance on circles and straight lines as the ultimate construction means admissible in geometry. Although there isn't any extant record about the explicit preference granted to straight lines and circles in Greek Antiquity, geometers from the Renaissance onwards often voiced their preference for Euclid's geometry as the only source of exact, geometrical knowledge.

Mathematical writers from the late Sixteenth and Seventeenth Centuries (Bos 2001, chapter 2) knew that cube duplication was solvable by employing more complex curves than straight lines and circles, such as the conic sections (Fig. 1.1). A similar result holds for the trisection of the angle. However, these mathematicians generally hesitated before admitting that solutions beyond the purview of Euclid's *Elements* could be truly geometrical. As we shall see below, even more prudence was shown in the face of the known solutions to the circle-squaring problem that appealed to transcendental curves like the quadratrix and the Archimedean spiral, curves considered ungeometrical *par excellence*.

Thus, it is not rare to encounter opinions in the literature of the time such as the following, expressed by François Viète:

> In geometry hyperbolas are not described in the way of true knowledge. Menaechmus doubled the cube by parabolas (...) but is the cube thereby doubled geometrically? Dinostratus squared the circle by the inordinate winding curve [i.e., the quadratrix] Archimedes by the

---

published together with an appendix on *Problematum quorundam illustrium constructionum*. The "illustrious problems" mentioned by Huygens include the duplication of the cube and the trisection of the angle. Huygens also devoted a long discussion to problems whose solutions depend upon the existence of a segment of given length, placed between two given curves (or between a curve and a straight line), which, extended, passes through a given point. This type of configuration is called, in scholarly literature, "neusis".

[3] Several studies have been devoted to the ancient and early modern tradition of the three classical problems. Among these, I shall address the reader to Knorr (1986) (ancient mathematics) and Bos (2001) (the Renaissance and the early modern era). For a survey of fallacious attempts to solve the quadrature of the circle, see Jacob (2005).

[4] Heath (1921, Vol. 1, Chapter 7) and Knorr (1986), Chapter 2 in particular.

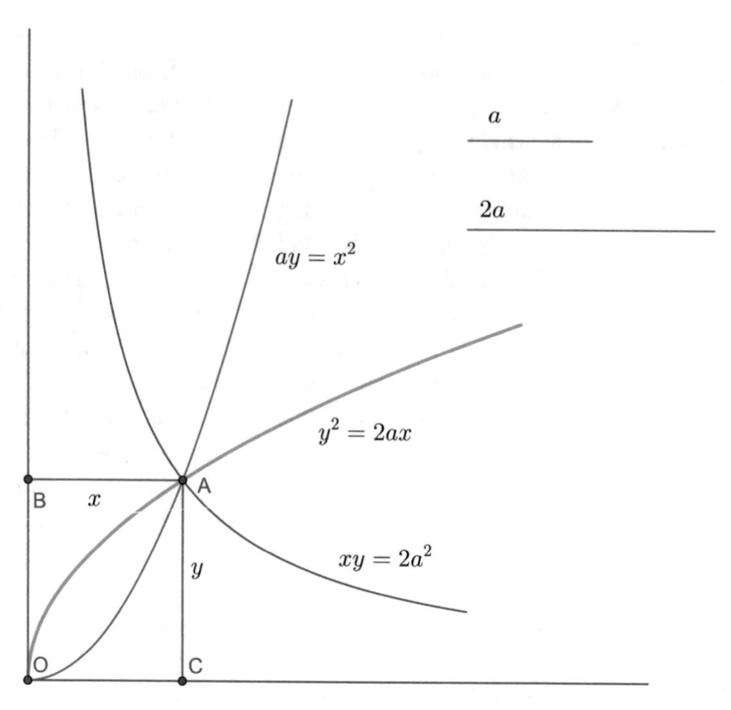

Fig. 1.1 According to the classical solution of Maenechmos, the duplication of a cube with side a (namely, the construction of a segment x, such that: $x^3 = 2a^3$) can be reduced to the problem of finding two mean proportionals x and y between segments a and $2a$ ($a : x = x : y = y : 2a$). The latter problem can then be solved by the intersection of a parabola and a hyperbola, for instance, or by the intersection of two parabolas. The solution to the duplication problem is the segment $x = BA$

> ordinate one [ie., the spiral], but is the circle thereby geometrically squared? No geometer would make that proposition. Euclid and all his disciples would raise in protest.[5]

On the other hand, it seems that, by the early Seventeenth Century, the unsolvability of the problems of duplicating the cube and trisecting the angle within Euclid's geometry was taken for granted, even if no impossibility proof had been given. Henk Bos stresses this very point, observing that:

> This impossibility (...) had been accepted as a matter of experience by classical Greek mathematicians, and most early modern mathematicians assumed, by experience or on authority, that the classical problems could not be constructed by straight lines and circles.[6]

[5]"Duplicavit cubum per parabolas Menechmus, per conchoidas Nicomedes, an igitur duplicatus est geometrice cubus? Quadravit circulum per volutam inordinatam Dinostratus, per ordinatam Archimedes, an igitur geometrice quadratus est circulus? Id vero nemo pronunciabit Geometra. Reclameret Euclides et tota Euclideorum schola." Quoted in Bos (2001, p. 178).

[6]Bos (2001, p. 24).

This conclusion matches well with the evidence on the cube duplication problem and the angular trisection,[7] but does not seem to hold for the circle-squaring problem as well. For example, Marin Mersenne, a lifelong friend and correspondent of Descartes, as well as a central figure in the French and European scientific milieu in the first half of the Seventeenth Century, refers to the solvability of the quadrature of the circle as an open question: "[the squaring of the circle] is extremely difficult, for one can find excellent geometers who claim that it is not possible to find a square whose surface is equal to that of the circle, and others who claim the opposite."[8]

Mersenne was thinking of the possibility or impossibility of solving this problem by circles and straight lines alone. As we shall see below, the issue about the solvability of the circle-squaring problem continued to be a disputable subject until the 1670s, even if, at that point, the reference framework was no longer represented by Euclid's geometry, but rather by Descartes'.

1.3 Impossibility Results in Classical Mathematics

Proving that a problem cannot be solved by prescribed means appears to be a conceptually different and more difficult task than searching for its solution. In fact, whereas, to solve a problem, it is sufficient to provide a construction, to prove its unsolvability, one should be able to survey the set of all imaginable constructions by certain means and check that not one of them produces the desired result. This is, of course, an impossible task for any geometer, because of the infinitude of possible constructions. For this reason, the unsolvability of the cube duplication, the trisection of an angle, and the quadrature of the circle by ruler and compass, or, equivalently, by straight-line-and-circle constructions, can be rigorously established only by using algebra to express arbitrary geometrical constructions as mathematical entities themselves. Such an approach became standardized only from the Nineteenth Century onwards, when the impossibility of the classical problems of geometry was also established in a rigorous way.[9]

The thesis underlying this book is that geometrical impossibility results were not completely neglected before the Nineteenth Century, but had an interesting history

---

[7]Several opinions on the unsolvability of these problems are cited in Bos (2001, Chapter 1).

[8]Quoted in Mancosu (1999, p. 79).

[9]The impossibility of solving the duplication of the cube and the trisection of a general angle by ruler and compass was first proven by Pierre Wanztel (1814–1848) in 1837. Wantzel's proof contains a mistake, discovered by Robin Hartshorne only recently (Lützen 2009, p. 383). The impossibility of squaring the circle was proved for the first time by Ferdinand Lindemann (1852–1939) in 1882, in a very strong way; in fact, he formulated the first proof of the transcendence of π, which implies the impossibility of squaring the circle with ruler and compass, and, more generally, with algebraic curves. For a mathematical treatment of the impossibility of the three classical, problems, I address the reader to Klein (1897) and the more recent (Jones et al. 1991).

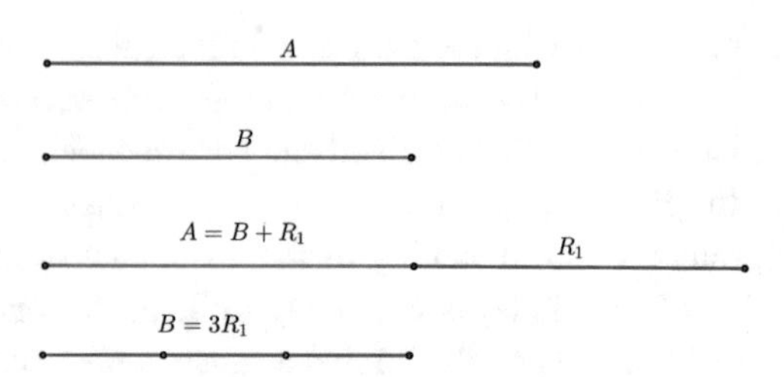

Fig. 1.2 The construction of a common measure between segments A and B (according to *Elements*, X, 3). B goes into A n_0 times, leaving a remainder R_1, less than B. The procedure can be repeated for B and R_1. As a result, it will leave a remainder R_2 less than R_1, etc. If and only if A and B are commensurable, the process will eventually terminate when the remainder is 0. The last step of the antyphairetic process will furnish the common measure between A and B

that contributed to establishing the idea that the classical problems of Antiquity were unsolvable quite before the discovery of rigorous impossibility proofs. My research builds on two recent studies by Lützen (2010, 2014), which argue convincingly that attempts at proving the unsolvability of the three classical problems and, particularly, the quadrature of the circle mathematically had begun in earnest by the middle of Seventeenth Century.

Even if I share this assumption in my study, I would like to remark that the very question about the possibility of solving a geometrical problem by prescribed methods might have been raised much earlier, already in the context of Greek geometry. One of the most outstanding results of Greek mathematics is indeed another famous impossibility result, namely, the discovery of mutually incommensurable segments. It is well known that Greek mathematicians had proven the existence of pairs of segments whose ratio is not in the ratio of counting numbers (or, as we would say in modern terminology, "natural numbers").[10] I would like to discuss here, instead, the lesser-known similarities between the phenomenon of incommensurability and the unsolvability of the three classical problems by straight-lines-and-circle constructions.

As Euclid's Book X recalls, two magnitudes are commensurable if they have a common measure. Let A and B be two such magnitudes A and B (think of them, for simplicity, as a pair of line segments) such that B is smaller than A. The problem of finding their common measure can be understood as a construction problem solvable by applying a geometrical version of Euclid's algorithm to find the greatest common divisor between two counting numbers. The Greek term for this method is "anthyphairesis," and it refers to the continual subtraction of the smaller of two unequal magnitudes from the greater (See Fig. 1.2).

If the magnitudes are mutually commensurable, as in our previous example, the process of repeated subtraction will eventually terminate, thus granting the construction of their common measure, and therefore the possibility of expressing their ratio as a ratio between numbers. However, the operation of repeated subtraction does not always yield a result

[10]For an informed historical discussion, Szabó (1978, pp. 214ff.).

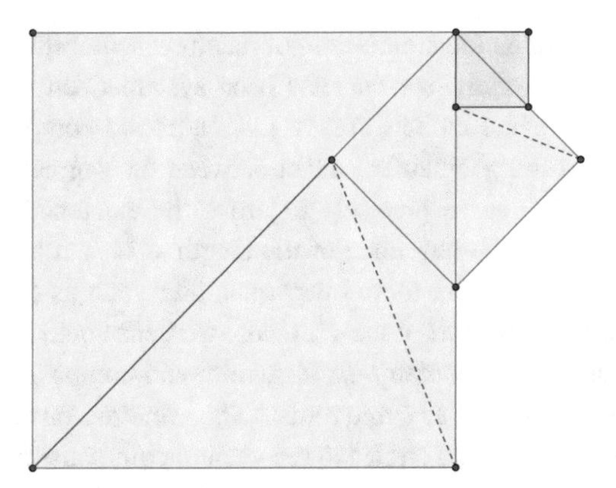

Fig. 1.3 The impossibility of constructing a common measure between the side and diagonal of a square. The well-known proof of the incommensurability between the side and the diagonal of a square establishes that the operation of repeated subtraction cannot be terminated, thus generating an infinite process, whose initial steps can be visualized in the figure (Szabó 1978, p. 210)

after a finite number of steps. This is the case if the given magnitudes are mutually incommensurable, as in the notorious example of the side and the diagonal of a square. For incommensurable magnitudes, the operation of repeated subtraction fails to terminate, generating an infinite process. A proof of incommensurability can be understood as a proof of the impossibility of constructing a ratio between the given magnitudes (e.g., the side and the diagonal) by the operation of repeated subtraction (see Fig. 1.3).

In pre-Euclidean geometry, the impossibility of measuring ratios between incommensurable magnitudes could have raised the theoretical problem of the existence of such ratios or of the measure of the length of certain segments, such as the diagonal of a square. The issue, on which we unfortunately possess scanty primary evidence from the Greek corpus, is less obvious than it might appear to us, since we take for granted an arithmetical conception of the continuum in which numbers such as $\sqrt{2}$, π and e exist on a par with natural numbers and with ratios between natural numbers. In his famous paper (Zeuthen 1896), Zeuthen explains how the incommensurability between the side and diagonal of a square was, on the contrary, a shocking result precisely because it proved that a measure of the diagonal of a square does not exist:

> For the Greeks who directly recognized as numbers only the positive integers, and indirectly also fractions as ratios of whole numbers, the proportion between incommensurable numbers was such that it could not at all be expressed by numbers. A number like $\sqrt{2}$ which, when multiplied by itself gives the number 2, simply did not exist for them.[11]

[11]Zeuthen (1896, p. 223): "Für die Griechen, die als Zahlen unmittelbar nur ganze Zahlen, indirect auch Brüche als Verhältnisse ganzer Zahlen, anerkannten, war das Verhältnisse incommensurabler Grössen ein solches, das sich überhaupt nicht durch Zahlen ausdrücken lies. Eine Zahl $\sqrt{2}$, die mit sich selbst multipliciert 2 giebt, existirte für sie einfach gar nicht."

Greek mathematics solved the conundrum of incommensurability by building a purely geometrical treatment of ratios, one that has been systematized in Books V, VI, X of the *Elements*. In particular, Zeuthen argues that ruler-and-compass constructions, by which we can construct the geometrical mean between the segments a and $2a$, namely, a segment of length $\sqrt{2}a$, serve precisely to prove the existence of incommensurable ratios. Furthermore, just as the existence of the length $\sqrt{2}a$ had to be proven through a geometrical construction, so were there other quantities such as the segment $x = \sqrt[3]{a}$, namely, the length of the side of the cube a^3, whose existence could not be granted except on the grounds of Euclidean geometry. In fact, ruler-and-compass constructions are not sufficient to secure the existence of cubic roots: "The same reasons [which brought about constructions by straight lines and circles to represent incommensurable ratios] must have led to the representation of a cubic root geometrically and the securing of its existence through a geometrical construction. Since [Greek geometers] did not succeed to do this through circle and straight line, they had to search for other constructions."[12] These other constructions could involve, for instance, conic sections or higher curves or methods that do not belong to plane Euclidean geometry.

Zeuthen takes these examples to illustrate the thesis that, in the tradition of Greek mathematics, geometrical constructions are to be understood as proofs of existence. According to him, the geometrical construction of an object x would establish that x does not contain any contradiction, and is thus logically possible. Since a proof that an object x is (constructively) possible logically entails the existence of x, Zeuthen concludes that one of the major goals of geometrical constructions and, *a fortiori*, of geometrical problems was to establish the existence of the sought-for objects, a goal motivated by theoretical concerns.[13]

This thesis has been criticized in Knorr (1983), particularly on the grounds that non-constructive existential assumptions are ubiquitous in the corpus of Greek mathematics. For instance, geometrical constructions cannot be existential proofs, Knorr argues, otherwise the existence of the solutions to the classical problems of antiquity would have been in doubt, which was never the case:

> If we could communicate to Euclid the findings of modern algebraic theory, showing that solutions of these problems are impossible under the restriction of the constructing techniques postulated in the Elements, would he conclude that the entities constructed in these problems in fact do not exist? We can well imagine that his response would be quite different: that these

[12]Zeuthen (1896, p. 224): "Eben dieselben Gruinde müssten dazu führen auch die Cubikwurzel geometrisch darzustellen and ihre Existenz durch geometrische Construction zu sichern. Da dies nicht dureh Kreis und Gerade gelang, musste man andere Constructionen aufsuchen," bracketed phrases added.

[13]Zeuthen (1896, p. 223): "die Construction mit dem dazu gehörigen Beweise für ihre Richtigkeit dazu diente, die Existenz desjenigen, was construirt werden sollte sicher zu stellen."

> entities obviously do exist; but that they merely cannot be constructed on the basis of the techniques he has postulated.[14]

In my opinion, Knorr's observation that the non-constructive existence of an angle measuring the third of a given angle, a cube being the double of a given cube, or a square having the same area of a circle was never questioned in Greek mathematics does not seem to contradict Zeuthen's thesis. To clarify the point, it may be useful to distinguish between two distinct senses in which a geometrical object can be said to exist: a first, "unconditional" sense, in which a certain mathematical entity exists independently from its construction, for instance, on the basis of a tacit assumption about continuity, and a "conditional sense," in which an entity exists in the background of a certain theory, such as Euclid's geometry. In the latter sense, to exist means to be "constructible within a theory or system."[15] As Knorr argues, the solutions to the classical problems of Antiquity did not necessarily raise concerns when taken as such, unconditionally or independently from a given theory of framework. On the other hand, we can question whether the problem of the existence of these entities mattered in the background of a theory such as Euclid's *Elements*. I think that Zeuthen's point is relevant precisely in connection with this distinction: to construct a geometrical object means, for him, to prove the conditional existence of the latter with respect to a given framework (for instance, Euclid's plane geometry). Therefore, if we could communicate to Euclid the unsolvability of the three classical problems, he might conclude that these entities exist unconditionally, but that they do not exist within the framework of the *Elements*.

In light of the distinction between conditional and unconditional existence, the existence of incommensurable magnitudes and the impossibility of constructing the solution to the three classical problems by circles and straight lines alone both appear to be instances of geometrical impossibility results of the same kind. In both cases, the impossibility depends on the limitation of certain prescribed methods, which are insufficient to prove that certain entities are constructible within a pre-assigned framework (i.e., the use of auxiliary constructions by straight lines and circles in the case of the classical problems, or the search for the common greatest measure between two segments by repeated subtraction in the case of incommensurability). Moreover, in all of these instances, supplementary methods are required to turn unsolvable problems into solvable ones. As Zeuthen explains, one needs constructions by straight lines and circles in order to represent incommensurable quantities such as $\sqrt{2}$, and one needs the use of conic sections, for instance, to represent ratios that are non-constructible by Euclidean means.

The analogy between the case of incommensurability and the classical problems is even more striking when we consider the circle-squaring problem. As we know, because of their

[14] Knorr (1983, p. 130).

[15] A similar point is made in Panza (2011, p. 43).

non-numerical and constructional approach to geometry, Ancient Greeks carried out the measuring of an area on purely geometrical grounds: by constructing a square equal to the figure to be measured. Therefore, unlike the case of polygons, whose area could be measured within the purview of Euclid's *Elements*,[16] Euclid's geometry lacked a method for determining the area of such a simple figure as the circle. Along similar lines, we can argue that Euclid's *Elements* lacked a method for establishing the length of circular arcs. Thus, the impossibility of squaring the circle by ruler-and-compass constructions is linked to the fact that the area of a circle (and the length of its circumference) cannot be measured within the *Elements*. Following Zeuthen's argument, we could also conclude that a square equal to a given circle simply does not exist in the framework of Euclid's *Elements*.

In conclusion, the circle-squaring problem, together with the problems of the duplication of the cube and the trisection of the angle, opened the general question about the conditional existence of certain entities within Euclid's plane geometry, in the same way as the incommensurability between a side and the diagonal of a square raised the issue of the existence of ratios between incommensurable magnitudes within pre-Euclidean geometry. Greek mathematicians had to face these challenges, and the extant records tell us that they tackled all of them similarly, by admitting new methods for construction: constructions by circles and straight lines in order to represent irrational magnitudes, or construction by higher means in order to represent cubic roots. The case of the area of the circle eventually required even more complex curves, such as the spiral or the quadratrix.

However, whereas an impossibility proof was devised for the case of incommensurability, this was not the case for the classical construction problems, as we know, until the surge of abstract methods from algebra. Yet, did the Greeks ever try to prove these impossibility results? The paucity of textual evidence leaves large room for speculation, but I think that there is also room for conjecture that Greek geometers devised ways to account for the impossibility of solving the classical problems, even if they were not able to produce a fully-fledged argument. In the next section, I will briefly illustrate an example of an ancient argument to prove the impossibility of solving the duplication of the cube and the trisection of the angle within Euclid's *Elements*.

1.4 Pappus on the Conditions of Solvability of Problems

According to the extant evidence, Greek mathematicians rarely discussed metamathematical issues, such as questions regarding classifications of problems, methods of solution, or the very issue of (un)solvability. A notable exception, which dates, however, from late Antiquity, is Pappus of Alexandria's *Mathematical Collection*. This is a miscellaneous work in 8 books, written during the early Fourth Century AD. Books 3 and 4, in particular, are dedicated to a survey of the classical problems and their solutions, which Pappus

[16]This can be done by applying *EL.* I, 45; II, 14.

presumably gathered from the tradition of pre-hellenistic and hellenistic mathematics. Although Pappus did not directly comment on the impossibility of solving the quadrature of the circle, he stated the unsolvability of the problems of the two mean proportionals and the trisection of the angle by plane methods, i.e., methods using only straight lines and circles. More generally, the theme of the solvability or unsolvability of geometrical problems is mentioned in the preface to the third book:

> On the other hand, he who puts forward a problem [in case he is ignorant and totally inexpert], even if he prescribes something which is somehow impossible to construct, is understandable and should not be blamed. Indeed, the task of the person who is searching is also to determine [*diorisai*] this: the possible and the impossible, and, if possible, when and how and in how many ways possible.[17]

Scrutinizing the limits of solvability of a problem was thus a fundamental task in the geometer's activity. Thanks to this examination, one could avoid erroneous results, such as in the case, discussed by Pappus at the outset of Book III of the *Collection*, of a flawed, plane (i.e., circles and straight lines) solution to the problem of duplicating the cube. This solution, attributed to an unnamed geometer (*quidam* in the Latin text), has been widely commented upon in the secondary literature.[18] I consider it here only insofar as it sparks Pappus' remark that plane constructions are improperly applied to solve a problem such as the insertion of two mean proportionals.

The Greek technical term for the analysis of the conditions of solvability of a given problem is "diorism." Several examples of diorisms can be found in the surviving corpus of ancient Greek mathematical texts, even in the *Elements*. A simple example can be found in proposition 22 from Book I of Euclid's *Elements*, which asks that a triangle be constructed given its three sides. The possibility of solving this problem obviously depends on the lengths of the given sides. More precisely, Euclid establishes, as a diorism, the necessary condition for the solvability of the problem that any side must be smaller than the sum of the remaining ones. Another, more elaborate example can be found in Proposition 27 of Book VI. In this proposition, Euclid proves a theorem on the application of areas. The theorem states that, of all of the parallelograms constructed on a given segment, whose defects are similar to a given parallelogram described on one half of the segment (Fig. 1.4), the greatest parallelogram is the one described on the other half. If we translate this problem into algebra, it becomes clear that this proposition also establishes the necessary condition under which a certain quadratic equation admits real solutions, hence it can be singled out as another example of diorism.

[17]In Bernard (2003, p. 119). I note that, in Hultsch's critical edition of Pappus' *Collection*, the bracketed expression is considered to be an interpolation.The Latin translation of Commandinus, current in the Sixteenth and Seventeenth centuries, contains the bracketed passage instead.

[18]For recent discussions of this case, see Bernard (2003) and Cuomo (2000), especially Chapter 5.

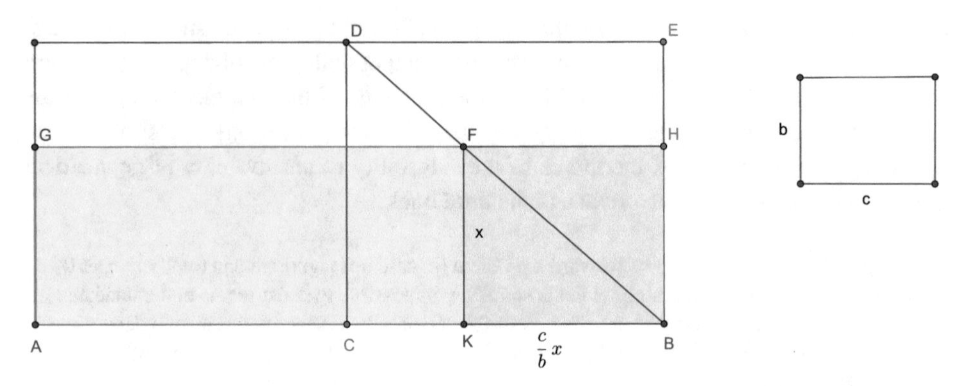

Fig. 1.4 The defect $FHBK$ is similar to the given rectangle of sides b and c. Let S be the area of the rectangle $GFKA$, and let $AB = a$. Hence: $ax - \frac{c}{b}x^2 = S$. Euclid proves that S cannot be greater than the area $DEBC$. This geometrical condition expresses the requirement that the discriminant of the equation $ax - \frac{c}{b}x^2 = S$ must be greater than 0

If a diorism is to be understood, generally speaking, as a statement that fixes the limit of solvability of a problem by establishing the necessary conditions for its solution, then a claim bearing on the unsolvability of the classical construction problems by straight lines and circles also falls into the same category, for the restriction of acceptable methods contributes to the necessary conditions for the solvability of the said problems. However, unlike the diorisms discussed by Euclid, proving the unsolvability of a problem by circle and straight lines appears to be beyond the technical resources of ancient geometry. What kind of argument, then, could ancient geometers deploy to formulate such diorisms?

In Book III of his *Collection*, Pappus claims that the insertion of two mean proportionals cannot be effectuated by relying on circles and straight lines alone, and promises a proof of this claim, as we can read in his own words: "the ancients already abandoned the hope that this problem [i.e., the insertion of two mean proportionals] could be solved by plane means, as I will myself prove, using their own statements."[19]

However, we do not find anything, in Pappus' discussion of the problem of the mean proportionals, resembling an impossibility proof, or at least any proof that we recognize as such. The only metamathematical remark given by Pappus is a classification of problems into three kinds, according to the curves used in their solutions:

> The ancients say that there are three kinds of problems in geometry, and some of them are called plane, others solid, and other linear. And so, those that can be solved through straight-line and circular-arc of a circle might fittingly be called plane. For the lines through which such problems are solved also have their coming-to-be on a plane. And those problems that are solved when one or more of the sections of a cone are employed for the discovery, these are called solid. For it is necessary for the construction to make use of the surfaces of solid

[19]Pappus (1876–1878, Vol. I, p. 45): "Hoc enim iam veteres per plana inveniri posse desperaverunt, sicut ipse appositis illorum sententiis demonstrabo."

> figures, I mean of conics. A third kind is left, which is called linear. For there are other lines besides those mentioned taken for the construction, as they have a coming-about that is more intricate and constrained, such as happen to be the spirals and quadratrices and cochloids and the cissoids (ivy-like), that have many and unexpected symptoms.[20]

An almost identical classification appears in Book IV of the *Collection*, in connection with a discussion on the nature of the problem of trisecting an angle,[21] and in Book VII, in connection with a classification of loci into plane, solid and linear.[22] In Greek mathematics, a locus, argues A. Jones, is "a definable object on which any point or line satisfying the conditions will be found, and such that any point that lies on the object will satisfy the conditions of the problem" (Pappus 1986, Vol. 2, p. 395). A plane problem is thus a problem whose solutions lie on a plane locus, since it can be found by the intersections of circles and straight lines, while the solutions of a solid problem, for instance, the duplication of the cube or the angle trisection, lie on a solid locus, i.e., the intersection between conic sections. Linear problems, the third genus singled out by Pappus, involved linear loci: this is the case of conchoids, or spirals, or any other curve more composed than conic sections, circles and straight lines. From this description, it appears that Pappus' classification is ultimately a classification of curves or loci.

More general constraints on the appropriate methods employable in solving mathematical problems are set by Pappus himself, in the course of the same discussion:

> Somehow, however, an error of the following sort seems to be not a small one for geometers, namely, when a plane problem is found by means of conics or of linear devices by someone, and summarily, whenever it is solved from a nonkindred kind.[23]

This requirement may be understood as a methodological principle of economy, that is to say, as a maxim imposed to solve a problem by the simplest possible means, circles and straight lines, whenever a problem is amenable to such a solution.[24] However, if we consider that the goal of the geometer was not simply to exhibit a solution of a problem via a construction, but also to prove the existence of this solution within a certain framework, then the above constraint becomes a fundamental condition for a problem to be solved. Instead of referring to a "framework," we could just as easily employ the locus-terminology. Thus, a problem solvable within the framework of the *Elements* is a problem solvable by plane loci, while a problem solvable within the framework of Apollonius' *Conics* is solvable by solid loci. In this sense, claiming that the solution of a problem

[20]Quoted in: http://web.calstatela.edu/faculty/hmendel/Ancient%20Mathematics/Pappus/Bookiii/Pappus.iii.1-27/Pappus.iii.1-27.html.

[21]Pappus (1876–1878, Vol. I. p. 273).

[22]Pappus (1986, Vol. 1, p. 106).

[23]Pappus (2010, p. 145) and Pappus (1876–1878, Vol. I, p. 271).

[24]See, for instance, the discussion in Knorr (1986, pp. 347–348).

belongs to a plane locus, when it belongs to a solid one, or vice versa, is not a mere matter of philosophical or methodological preferences: it is an error with respect to the task set by the problem itself, namely to prove the conditional existence of a certain geometrical entity within a given framework or theory.

Despite its methodological interest, it must be stressed that Pappus' considerations on the division of problems and the correlative error consisting in solving a problem by the non-kindred type of locus is not to be found elsewhere in the corpus of Greek mathematics.[25] Given its unique occurrence, Pape (2017) has argued that Pappus' tripartite classification appears as an argument deployed within a particular context, with the aim of disproving a false, plane solution to a solid problem, such as the one discussed at the beginning of Pappus' Book III, which regards the insertion of two mean proportionals.

Pappus' line of argument could thus have proceeded in this way. The nature of certain problems, such as the cube duplication or the trisection of the angle, is solid, as ascertained by a well-established tradition of research. This statement works as a kind of diorism, because it provides the limit of solvability of the above-mentioned problems, and establishes, as a necessary condition, that the solution to such problems as the cube-duplication and the angle trisection exists only among solid loci.[26] Any attempt to solve this problem that violates this condition is therefore doomed to failure, as in the case of the "quidam" solution. Here, the tripartite classification can be seen as an attempt by Pappus to provide a stystematization of these historical facts, and also to provide a rational justification for the fault in the "quidam" solution.

This argument is hardly recognizable as an impossibility proof by contemporary readers. In fact, it rests more on an extramathematical rationale based on the tradition of geometry than on mathematical reasoning. However, in the absence of formal methods for tackling impossibility results, it was probably the only kind of argument available to Pappus to argue for the unsolvability of the cube duplication or the angle trisection by circles and straight lines.

To conclude, while we earlier stressed the analogy, in the context of ancient and late-ancient Greek mathematics, between the incommensurability between the side and diagonal and the impossibility of solving the classical problems by ruler and compass, we must now stress the differences. The main divergence appears to occur when it comes to the proof of these results: while the Greeks had possessed, for the incommensurability case, a fully-fledged mathematical proof since pre-Euclidean times, they presumably lacked a mathematical proof for the unsolvability of the classical, construction problems. Their only impossibility argument, as it appears, revolved around extramathematical considerations.

This would also be their legacy to future generations, and, particularly, the starting point of their early-modern inheritors. As we shall see, in fact, Pappus' argument was

[25]Pappus (1986, vol. 2, p. 530).

[26]For a different interpretation of Pappus' claim in a stronger, metaphysical sense of a "homogeneity requirement," see Sefrin Weis' commentary in Pappus (2010, p. 274).

revived by Descartes in the Seventeenth Century and crucially modified in order to obtain a mathematical proof of impossibility.

1.5 On the Impossibility of the Classical Problems in the Seventeenth Century

Pappus' classification exerted a large influence on early modern geometers, who saw it as the royal road upon which the Greeks ordered their own subject matter via a hierarchy of solving methods. In particular, this classification was evoked or paraphrased when considerations about the solvability or unsolvability of problems by given means were at stake. A suggestive case is offered by Descartes' mathematical masterpiece, the *Géométrie* (1637). As Lützen argues in his (Lützen 2009), it is not just by chance that Pappus' *Collection* and Descartes' *Géométrie* both deal with impossibility results in geometry. Both works, in fact, contain substantial reflections about mathematics and its method (Lützen 2009, p. 7), and a discourse about the impossibility of solving problems could fit within such reflections.

As a more precise confirmation of Lützen's claim, we can stress an analogy between Pappus's introduction of his tripartite classification in connection with a discussion about a flawed solution to the problem of the two mean proportionals and Descartes' introduction of a classification of problems, together with considerations on mathematical impossibility, within a methodological discussion about the kind of geometrical errors that emerge when one does not proceed according to the rules of the method. This discussion takes place with particular prominence in the third book of the *Géométrie*, which is also largely devoted to the solution of polynomial equations.[27]

In Descartes' classification of curves and problems, algebra indeed plays a fundamental role. More generally, algebra is at the core of Descartes' problem-solving method, in which: "all problems which present themselves to geometers reduce to a single type, namely, to the question of finding the value of the roots of an [finite, algebraic] equation."[28] The salient epistemological features of Descartes' method are all included in the outline given above: the method is strictly finitist, in the sense that it does not, and cannot, deal with "infinitary objects" like infinitesimals or indivisibles, and it is constructivist, in the sense that geometrical knowledge can be acquired only via an explicit geometrical construction involving legitimate (i.e., algebraic) curves.[29]

[27] Cf. Descartes (1897–1913, vol. 6, p. 288, 370).

[28] Descartes (1952, p. 216), with minor modifications.

[29] Giorgio Israel clearly points out two important consequences of the Cartesian finitism and constructivism in geometry: "any form of reasoning *ab absurdo* is excluded in Cartesian mathematics; moreover, the entities all have to be constructible, which makes it impossible to define them in a conventional or axiomatic way. Furthermore, the admissible deductive chains must be finite; consequently, also the rudimentary forms of inductive reasoning in Descartes' work differentiate

By a mix of geometrical and algebraic considerations, Descartes thus reshaped Pappus' distinction into three kinds of problem and stressed the importance of avoiding errors stemming from assigning a problem to the wrong class, with an emphasis we do not find in Pappus' *Collection*. In this regard, Descartes articulated, in the third book of the *Géométrie*, a "norm" according to which it is an error in geometry to try to solve a problem by means that are either overly simple or overly complex.[30] This norm is summarized by Descartes himself in just one phrase:

> We have to be careful, when solving a problem, that we must never take lines of too complex a kind, except when it is impossible to accomplish what is required with lines of a simpler kind.[31]

Descartes had in mind specific examples, such as the solution of the two mean proportionals problem by conic sections, as per in Maenechmus' solution, in contrast to the solution of the same problem using the intersection between a circle and a more complex curve engendered by a special tracing device. The complexity of curves, on which a flawless solution depends, is, for Descartes, determined by the degree of the equation corresponding to the curve.

In connection with the question of the avoidance of error in solving problems, Descartes also discussed the central question as to why "solid problems cannot be constructed without conic sections, nor the more composed ones by more complex curves." Firstly, Descartes made it clear that attributing the correct level to a problem depends on algebra:

> If it is remembered that in the method I use all problems which present themselves to geometers reduce to a single type, namely, to the question of finding the values of the roots of an equation, it will be clear that a list can be made of all the ways of finding the roots, and that it will then be easy to prove our method the simplest and most general.[32]

---

from modern inductive reasoning which, by means of a finite number of steps, makes it possible to pass from the finite to the infinite." In Israel (1997, p. 4).

[30]Descartes (1897–1913, vol. 6, p. 370).

[31]Descartes (1897–1913, vol. 1, p. 460): "il faut prendre garde, aux solutions de 'Problèmes, qu'on n'y doit jamais employer des lignes courbes d'un genre composé, que lorsqu'il est impossible de faire ce qui est requis avec des lignes de plus simple genre." So Descartes wrote to an unknown correspondent, possibly Godefroid de Haestrecht.

[32]Descartes (1952, p. 216). Cf. the original: "Il est vray que je n'ay pas encore dit sur quelle raison je me fonde, pour oser ainsy assurer si une chose est possible, ou ne l'est pas. Mais, si on prend garde comment, par la methode dont ie me sers, tout ce qui tombe sur la consideration des Geometres, se reduit a un mesme genre de Problesmes, qui est de chercher la valeur des racines de quelque Equation, on iugera bien qu'il n'est pas malaysé de faire un denombrement de toutes les voyes par lesquelles on les peut trouver, qui soit suffisant pour demonstrer qu'on a choisi la plus generale et la plus simple" (Descartes 1897–1913, Vol. 6, p. 475).

But, when he turned to examine the case of solid problems, a surprising switch from algebra to geometry took place, and he offered a fully synthetic impossibility argument instead. Descartes generalized the same argumentational scheme to problems "one degree higher than solids," which presuppose the invention of four mean proportions, or the division of an angle into five congruent parts[33] in order to prove that these problems cannot be solved by conic sections, but require higher curves, like the Cartesian parabola.[34]

Descartes' impossibility argument can be summarised as follows.[35] Since the curvature of the circle depends on one "simple relation" (*un simple rapport*, to be understood as the distance from the center of all of the points on the circumference), this curve can be used to construct, at most, one point between the extremes of a segment or arc. Since the curvature of a conic section depends on two "things" (*choses*) or relations, left unspecified by Descartes, it can be employed in order to determine, at most, two points between two (given) extremities. In the case of the conic sections, Descartes could be referring to the distances of all of the points of a conic from each of the foci. In short, then, the circle cannot be employed to construct problems that require the determination of two points, namely, the trisection of the angle or the insertion of two mean proportionals.[36] Descartes' presentation is slightly more detailed, but does not dispel the impression, evident at least to a present-day reader, that his argument claiming that solid problems could not be solved by plane means is fairly obscure and certainly questionable in some of its assumptions. In spite of this fact, it did not fall into oblivion. Later attempts made in the course of the Seventeenth and Eighteenth Centuries to prove the impossibility of solving the two classical construction problems consisted largely in working out the implications of the argument offered by Descartes in the *Géométrie*.[37]

All in all, Descartes' *Géométrie* stood as a cornerstone in the early modern history of impossibility in geometry. Firstly, impossibility results were functional to Descartes' methodology, according to which one had to take into account a precise hierarchy of methods when solving a problem. In this context, establishing that a problem could not be solved by certain curves enabled the geometer to rule out overly simple solutions for the question at hand. Secondly, Descartes also established a connection between analytical or algebraic reasoning and impossibility theorems and proofs.

The reasons why Descartes tried to prove impossibility in a purely geometrical way are, from our viewpoint, striking.[38] However, one might consider that the algebraic techniques

---

[33]Descartes (1897–1913, Vol. 6, p. 476).

[34]"La ligne courbe qui se descrit par l'intersection d'une Parabole et d'une ligne droite ... car j'ose assurer qu'il n'y a de plus simple en la nature, qui puisse servir a ce mesme effet" (Descartes 1897–1913, Vol. 6, p. 476).

[35]Descartes (1897–1913, Vol. 6, p. 476).

[36]Cf. Bos (2001, pp. 379–380) and Lützen (2009, pp. 22–23).

[37]See Lützen (2009, pp. 28ff.).

[38]See also Lützen (2010, p. 22).

for carrying out such proofs in all generality were not available at the time of Descartes. Moreover, algebra was, overall, a method of discovery, that is to say, a method for finding solutions to problems, rather than establishing proofs of theorems. In light of these considerations, the choice to search for a geometrical proof of impossibility might have come naturally to Descartes. Moreover, those geometrical impossibility proofs commented upon above could be easily generalized to provide arguments for the unsolvability by ruler and compass of higher problems. For instance, one might argue that the division of an angle into five parts could not be effectuated by the intersections of circles and straight lines for about the same reasons that, according to Descartes, the unsolvability of the trisection is not solvable by Euclidean methods: the number of points required in the sought-for configuration exceeds the number of possible intersections between a straight line and a circle. This type of "enumerative" argument, even if lacking cogency, may be easily extended to cover more and more complex kinds of problem (such as the division of the angle into a higher number of parts).

In spite of Descartes' avoidance of algebra in his impossibility arguments, Descartes' algebraic technique in problem-solving was later employed to prove impossibility theorems, especially in connection with the quadrature of the circle. Indeed, as we shall see, both Gregory's and Leibniz's impossibility proofs depended on algebraic arguments.

1.6 The Problem of Squaring the Circle Until Descartes

Beyond the problem of squaring polygons, whose general solution was within the purview of Euclid's Elements,[39] Greek geometers obtained important results concerning the exact quadratures of curvilinear surfaces as well. A well-known example is the quadrature of the parabola found by Archimedes. Using two different methods, a mechanical proof and a formal indirect proof, Archimedes proved that the area of a parabolic segment P (namely, the figure bounded by a parabolic arc and having as its basis a chord of the parabola) is $\frac{4}{3}$ of the triangle T_0 having the same basis and same height as the parabolic segment.[40] This is not *per se* a construction, but gives us an easy rule for constructing a square equal to the given parabolic segment using only straight lines and circles.

However, no similar, exact results were found for the quadrature of the circle. According to a sketchy, although globally correct, survey given in Tropfke's classical manual, Greek geometers pursued three general directions of research regarding the circle-squaring problem.[41] The first one consisted in trying to construct a square exactly equal to a given circle by ruler and compass only. Such an attempt is obviously bound to fail, since the quadrature of the circle is not a plane problem. I do not have any knowledge of such

[39]This can be done by applying propositions I, 45 and II, 14 of Euclid's *Elements*.

[40]Archimedes (1881, Vol. 2, pp. 293ff.); for a translation into English: Heath (1897, p. 233).

[41]Tropfke (1903, p. 110).

flawed attempts to solve the circle-squaring problem in Antiquity, but it would not not be so astonishing to discover that they had occurred, as they did in the Renaissance or early-modern period. Tropfke could also be thinking of certain alleged solutions criticized by Aristotle in several passages of his corpus, such as the purported quadrature of Bryson.[42] From the sole surviving, indirect evidence, and from Heath's commentary, we could reconstruct Bryson's argument thus: as a circle is smaller than any circumscribed polygons and larger than any inscribed one, one could start from a pair of in- and circumscribed squares and, by continually halving their sides by ruler and compass, obtain a sequence of in- and circumscribed regular polygons. The sequence could be so continued until we obtain a pair of polygons, I_n and C_n, differing so little in area that, if another intermediate polygon K can be described so that $I_n < K < C_n$, K will have the same area of the sought-for circle. Since it is a matter of elementary geometry to construct a square equal to a given polygon, the quadrature of the circle would be solved as soon as K is constructed. However, according to the extant indirect evidence, Bryson fails to explain precisely how K would be found. In as much no recipe is given to construct a polygon equal to the circle, Bryson's argument would work more as an argument that claims the unconditional existence of a solution to the circle-squaring problem, rather than a proof that such a solution is available, for instance, within Euclid's geometry.

According to Tropfke, ancient geometers always followed a second route to solve the circle-squaring problem using mechanical curves such as the quadratrix or the Archimedean spiral, generated by special combinations of motions (both curves are described below, in Figs. 1.5 and 1.6).

However, the spiral and the quadratrix do not directly solve the quadrature of the circle, but only the problem of rectifying its circumference. As we know, these problems are equivalent. This non-trivial result was first proven in Archimedes' *Measurement of the Circle*. Thus, the role of mechanical curves in solving the quadrature of the circle was presumably recognized only after Archimedes' result.

In the short treatise on measuring the circumference, part of a longer, lost work by Archimedes established that:

> The area of any circle is equal to a right-angled triangle in which one of the sides about the right angle is equal to the radius, and the base [i.e., the other sides about the right angle] to the circumference [of the circle].[43]

That is to say, the ratio between the circle and the square built on its radius (we may call this ratio π_1, which is equal to the ratio between the circumference and the diameter itself (that is, π_2; hence $\pi_1 = \pi_2 = \pi$)). Therefore, in order to determine the area of a circle with given radius, it is sufficient to know the length of its circumference.

[42] Heath (1921, pp. 322–323, Vol. 1).

[43] Dijksterhuis (1987, p. 222).

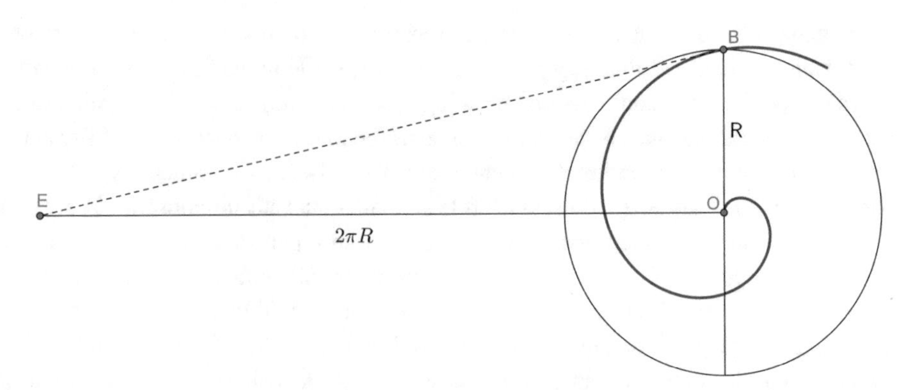

Fig. 1.5 According to Archimedes' definition, the spiral is generated by the uniform motion of a point along a segment while the segment simultaneously rotates along one of its fixed ends. As can be easily ascertained using calculus, the subtangent EO to the spiral $r = R\frac{\theta}{2\pi}$ equals the length of the circumference with radius R. The same result is proven, using a double reductio argument, in Archimedes' treatise *On Spirals*

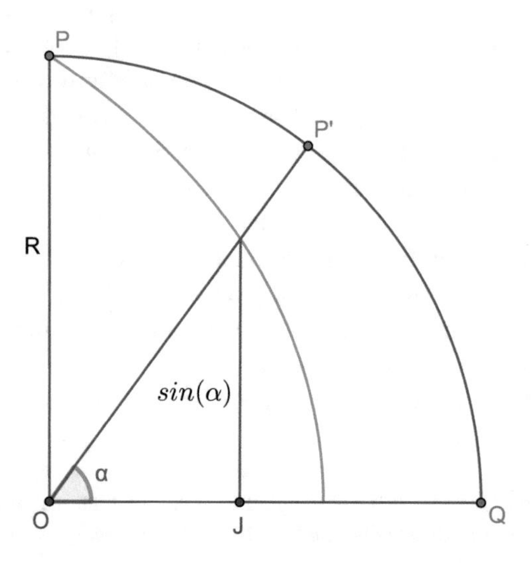

Fig. 1.6 According to Pappus, the quadratrix is the curve traced by the point of intersection between the uniformly rotating radius OP and a segment perpendicular to PO, which translates along it with uniform motion. Both movements have to be synchronized in order to trace the curve. As per its definition, in a quadratrix, we have that: $R : sin(\alpha) = \frac{\pi}{2} : \alpha$. This proportion easily leads to the polar equation of the curve, namely, $r = \frac{2sin(\alpha)}{\pi R\alpha}$. Therefore, when the rotating radius tends to $\alpha = 0$, the curve cuts a segment $OJ = \frac{2}{\pi R}$ on the horizontal axis

Archimedes' original proof is furnished via an indirect method known in the scholarly literature as the "compression method."[44] This technique can be described in its generality as follows. The magnitude to be calculated, in our special example, the area of the circle, is squeezed between a monotonically increasing polygonal sequence I_n and a monotonically decreasing sequence C_n. Archimedes' proof also holds that the difference $C_n - I_n$ can be made smaller than any assigned magnitude ϵ. A magnitude K is then found, such that, for every n, $I_n < K < C_n$. In the end, Archimedes proves, via a double *reductio ad absurdum*, that K is equivalent to the area of the circle.

To find K in the special case of the circle-squaring problem, one may start from the following facts, which are implicit in the extant version of the *Measurement of the Circle*:

- The perimeters of every inscribed polygon are smaller than, and those of every circumscribed polygon greater than, the circumference of the circle.
- The in-radii of the polygons in- and circumscribed to a given circle are respectively less than and equal to the radius of the circle.
- The area of a regular polygon is equal to the rectangle formed by one-half of its perimeter and its in-radius.

On the strength of these assumptions, it can be inferred that the rectangle formed by one-half of the circumference of a circle C and its radius is greater than the area of every regular polygon inscribed in C, and smaller than the area of every regular polygon circumscribed to it. Archimedes then continues by furnishing a rigorous indirect proof that the rectangle formed by one-half of the circumference of a circle C and its radius has the same area of the circle. The same result can be visualized using an intuitive, heuristic reasoning based on dissecting the circle into countless sectors and then by pasting them so as to form a rectangle, as is done in the intuitive argument depicted in Fig. 1.7. This operation of cutting and pasting will not precisely yield a rectangle if the number of triangular sectors is small, but becomes more and more exact as the number of sectors increases.

It should be highlighted that Archimedes did not furnish a constructive solution to the circle-squaring problem. However, as a follow-up to the above result, he devised an ingenious method for approximating the length of the circumference by suitably truncating the sequences of in- and circumscribed regular polygons that "compress" the circle. In this way, Archimedes could produce the very accurate numerical estimate $3\frac{1}{7} < \pi < 3\frac{10}{71}$. This approach represents a third direction of research into the circle-squaring problem, which does not consist in finding a construction properly, but rather an approximate computation of the area or the circumference of the circle. Archimedes' method of approximation remained, until well into the Seventeenth Century, when reasoning with infinitesimals

[44] Dijksterhuis (1987, pp. 222ff.).

Fig. 1.7 A visual illustration of Archimedes' proof that $A = \frac{1}{2}cr$ (A is the area of the circle, c its circumference and r the radius), Heuristic arguments of this kind represented expedient alternatives to the lenghtier, formal proof given by Archimedes and were frequent in the middle ages (cf. Smeur 1970, p. 254), but they might originally have been used by Archimedes in order to discover the relation between the length of the circumference and the area of a circle. Image created with GeoGebra, by Linda Fahlberg-Stojanovska

and the application of infinite series offered new insights into the nature of π, the only successful method to tackle the century-old problem of squaring the circle. As an example of the fortune of this method, we can remember how, still in 1654, Huygens had published a successful work on the measurement of the circumference, *De circuli magnitudine inventa*, in a perfect Archimedean spirit. In fact, he obtained more accurate bounds for π than Archimedes, still adhering to the fundamental idea that the area of the circle can be "compressed" by sequences of in- and circumscribed polygons.

However, even if this approach could lead to very accurate estimates, it failed to satisfy the criteria of exactness imposed on geometrical constructions. Certainly, such inadequacy was already noticed in late Antiquity. Ammonius, a philosopher who composed a commentary on Aristotle's *Categories* between the Fifth and Sixth Centuries AD, explained how the quadrature of the circle was still an open problem at his time:

> Having raised a square equal to a rectilinear figure, geometers also sought, if possible, to find a square equal to a given circle. Many geometers - including the greatest ones - looked for it, but did not find it. Only the divine Archimedes discovered anything at all close, but so far the exact solution had not been discovered. Indeed it may be impossible.[45]

[45]Quoted in Knorr (1986, p. 362), with slight modifications.

Although he was educated as a philosopher, Ammonius was conversant with astronomy and mathematics, so that his testimony can be taken seriously (Keyser 2017). By remarking that an "exact" solution has not been discovered yet, the commentator could have meant a solution obtained by Euclidean means. This also explains why Archimedes is mentioned as the one who has discovered anything close to a solution: his polygonal procedure for determining the length of the circumference relies on the constructive means of Euclidean geometry. Although this method does not provide an exact rectification, it does provide an infinite procedure, which can be truncated after arbitrarily many steps to offer an approximate estimate of π in a finite number of steps. From the above passage, we also understand that the solvability of the circle-squaring problem was by no means a decided issue for authors from late Antiquity. This would put the quadrature of the circle on a different level than the other classical problems of geometry, whose non-plane nature was accepted and sometimes justified metamathematically, as in the above-mentioned case of Pappus.

The uncertainty of the ancient or late-ancient commentators regarding the possibility of solving the circle-squaring problem resonates well with the opinion that would hold for centuries afterwards, that of the Arabic mathematician Ibn al-Haytham:

> Many philosophers have believed that the area of a circle cannot be equal to the area of a square limited by straight lines. This notion has been revisited many times in their dialogues and controversies, but we have discovered no early or modern work containing a polygonal figure exactly equal to the area of a circle. Archimedes made use of a certain approximation in his work on the measurement of the circle, and this latter notion is among those that have reinforced the opinion of the philosophers in their conviction.[46]

This treatise was written in the Eleventh Century of our era. While it is unclear as to which (if any) particular controversies Ibn al-Haytham is referring in the passage, it is, on the other hand, apparent that, for the Arabic mathematician, no shared solution had been given, at the time of his writing, to the circle-squaring problem. Archimedes' approximate solution is the only ancient result quoted, and it is used to justify the "philosophical" belief that an exact solution to the problem could be found. Later on in the same text, al-Haytham claims that the circle-squaring problem can be proven to be "possible" on a sound epistemological basis. Let us recall that this argument is addressed to philosophers rather than mathematicians, and the special choice of the audience may justify the type of justification presented (see Fig. 1.8 for a summary), which does not provide a geometrical construction, but only an argument for the existence of a solution, that is to say, a proof of the unconditional existence of a square equal to a given circle.

Yet, an addendum to the text written by an objector (Álī ibn Riḍwān or al-Sumaysāṭā, c. 988-c. 1061) points out precisely the apparent failure of al-Haytham's proof of possibility:

[46] *Treatise on the Quadrature of a Circle*, in Rashed (2013, p. 99).

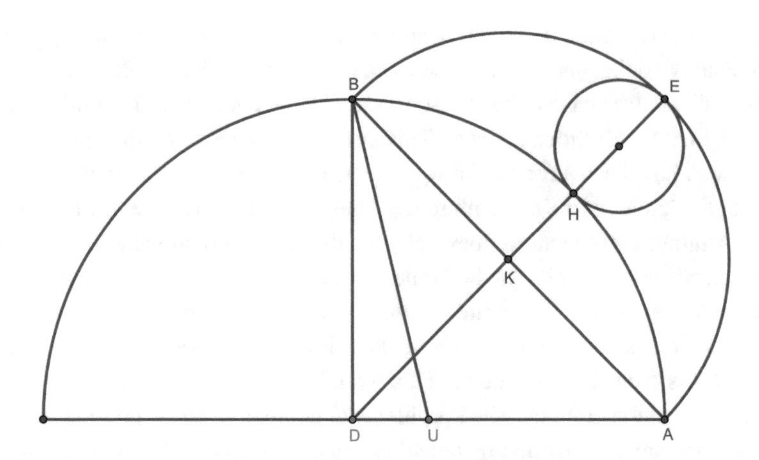

Fig. 1.8 We summarize here Ibn al-Haytham's argument to show that the quadrature of the circle is solvable. This argument depends on a previous result obtained by Al-Haytham, namely, the proof that the lune $BHAE$ has same area of the triangle BDA. A circle with diameter HE is constructed as in the figure. Since the circle HE is contained in the lune $BHAE$, it is smaller than it (by Euclid's common notion 5) and will have a certain unknown ratio k with the lune itself. By hypothesis, let $\frac{BHAE}{c(HE)} = \frac{AD}{DU} = k$, where $c(HE)$ is the circle with diameter HE and U is a point chosen arbitrarily on AD. Next, we have, by elementary geometry, $\frac{BHAE}{c(HE)} = \frac{AD}{DU} = \frac{BDA}{BDU}$. Since the lune $BHAE$ is equivalent to BDA, the circle with diameter HE will be equivalent to the triangle BDU, which can be transformed into an equivalent square. al-Haytham's justification depends, eventually, on the existence of the ratio k. However, this ratio is itself an algebraic function of π

> To my eyes he [Ibn al-Haytham] has done nothing in this treatise, as that which was sought was to construct a square equal to the circle. Whether this is possible or not in the divine knowledge does not help us in finding that which is sought. In saying that this is possible without our having the ability [to construct it] adds nothing to the belief held by the earlier authors ... If the knowledge of something is inaccessible to us, then that thing is inaccessible, and our conviction that knowledge of it is possible is of no use whatsoever.[47]

In short, geometrical knowledge is acquired by construction, but a construction of the circle-squaring problem (which amounts to a construction of the constant π) has not been provided by anyone. To conclude this overview, it is worth pointing out that, as Mersenne's words quoted in Sect. 1.2 suggest, until the first half of the Seventeenth Century, geometers had not advanced much on the issue of the exact solvability of the circle-squaring problem. This lack of progress was such that either Ammonius' words or the critique by al-Haytham's anonymous objector could have been shared by an early Seventeenth Century scholar, like Mersenne himself, without much quibble.

The distinctions between existence and construction on one hand and between exact and approximate solutions on the other were also clear to early-modern geometers,

[47] Quoted in Rashed (2013, p. 105).

who often relegated approximate solutions to practical geometry or mechanics, while exact solutions were the prerogative of geometry proper. Along these lines, Leibniz still distinguished, well into the Seventeenth Century, between the practical goal of finding better computations for the constant π and the task of finding an exact solution to the circle-squaring problem:

> Archimedes, inscribing and circumscribing polygons to the circle, because the latter is greater than the inscribed polygons and smaller than the circumscribed one, shows a way of presenting the limits, between which the circle must fall, or of showing the approximations: clearly the ratio of the circumference to the diameter is greater than 3 to 1 or than 21 to 7, and less than 22 to 7 ... Truly approximations of this kind, even if they are useful in practical Geometry, yet show nothing, which may satisfy the mind in great need of the truth, unless a progression of such numbers being considered to infinity may be found.[48]

For many geometers after the publication of Descartes' *Géométrie*, the paradigm of a solution that could appeal to a mind eager for truth and not merely pleased by practical matters was no more given by ruler-and-compass constructions than it had been by algebraic curves. Cartesian geometry proved to be of very little help when it came to the squaring of the circle. Descartes himself denied that the circle could be algebraically squared, but did not really give any argument to support his views. He simply took for granted a more general claim that the exact ratio between straight and curvilinear lines is unknowable to men:

> The ratios between straight and curved lines are not known, and I believe cannot be discovered by human minds, and therefore no conclusion based upon such ratios can be accepted as rigorous and exact.[49]

According to the conjecture advanced in Baron (1969, pp. 223–228) and endorsed by Bos (1981, p. 314), such an unjustified, almost dogmatic belief might be traced back to an ancient classification of lines whose earliest known occurrence can be found in Aristotle's *Physics*. In the passage in question, Aristotle, probably relying on contemporary mathematical wisdom, assumes that straight and curvilinear magnitudes belong to different kinds, hence they cannot be compared one to the other. Averroes explains the Peripatetic view on the grounds that straight lines and arcs cannot be made to coincide with each other, as explained in a commentary to Aristotle's *Physics* (written in the first half of the Twelfth

[48]"Archimedes quidem Polygona Circulo inscribens & circumscribens, quoniam major est inscriptis, & minor circumscriptis, modum ostendit, exhibendi limites intra quos circul[u]s cadat, sive exhibendi appropinquationes: esse scilicet rationem circumferentiae ad diametrum, majorem quam 3 ad 1, seu quam 21 ad 7, & minorem quam 22 ad 7... Verum hujusmodi Appropinquationes, etsi in Geometria practica utiles, nihil tamen exhibent, quod menti satisfaciat, avidae veritatis, nisi progressio talium numerorum in infinitum continuandorum reperiatur" (Leibniz 2011, pp. 8–9).

[49]Descartes (1897–1913, Vol. 6, p. 412).

Century) that enjoyed a large circulation in Latin translations during the Renaissance. As a consequence, there is no geometrical criterion to establish an order relation between them, whereas it is possible to compare segments with segments, as well as arcs with other arcs, provided they lie on the same circle.[50]

Thus, if we follow Averroes' reading of Aristotle closely, it would be inconceivable for rectilinear and curvilinear magnitudes to have a mutual ratio, as, according to definition 4 of *Elements*'s Book V: "magnitudes are said to have a ratio to one another which are capable, when multiplied, of exceeding one another." To rectify an arc would be an impossible operation, just like measuring a volume by either a surface or a linear magnitude.

This view encountered many sound objections in later history, coming, not surprisingly, from the Archimedean tradition.[51] The arguments of Bryson or Ibn al-Haytham, quoted above, either postulated or argued for the existence of a solution to the circle-squaring problem, albeit in a non-constructive way. A possible goal of such arguments may have been to refute the hypothesis that circles and polygons are heterogeneous quantities, a hypothesis that, if true, would have made it a priori impossible to square the circle.

Moreover, ancient and Medieval commentators felt bound to supplement the proof of Theorem 1 of Archimedes' *Measurement of the Circle* with a postulate stating that one could actually produce a straight line equal to a circle.[52]

Eutocius, a commentator of Archimedes, strongly affirms that the existence of a straight line equal to a circumference is an obvious matter, even if its actual construction had not been found out yet:

> For it is somehow clear to everyone that the circumference of the circle is some magnitude, I believe, and this is among those extended in one [sc. dimension] while the straight line is of the same kind. Even if it seemed not yet possible to produce a straight line equal to the circumference of the circle, nevertheless, the fact that there exists some straight line by nature equal to it is deemed by no one to be a matter of investigation.[53]

Insisting on the obviousness of this fact, the medieval *Corpus Christi* version of the *Measurement of the Circle* incorporated the claim of the existence of a straight line equal to a circle, and, more generally, the concern for the comparability of straight and circular

[50] I shall quote Bagolinus' Latin translation: "Non est proportionalitas secundum veritatem inter lineam rectam et circularem ...et intendebat per hoc, quod impossibile est de quantitatibus esse aequales nisi rectas tantum aut circulares tantum, scilicet quae sunt ejusdem speciei, cum istae sibi superponantur; et ideo dicimus, quod quantitates curvae non aequabuntur nisi sint ejusdem circuli," in Hofmann (1942, p. 6).

[51] For an overview, see: Molland (1991, p. 192).

[52] The postulate is explicit in Eutocius' commentary (Cf. Archimedes 1881, Vol. 3, p. 267), and in two Medieval commentaries, the Cambridge and the Corpus Christi manuscripts of *the Dimensio Circuli* (Clagett 1964, p. 68, 170, pp. 382ff., pp. 414ff.).

[53] Quoted in Knorr (1986, p. 362).

segments, by interpolating the original Archimedean text with three postulates, "known per se and recognized by anyone," the anonymous commentator remarks:

> The first of the three postulates is that an arc is greater than [its] chord. The second of the postulates is that a curved line be equal to a straight line. The third of the postulates is as follows: any curved line sharing the two termini or a circumferential arc and including it in the direction of the convexity of the arc, is greater than the arc.[54]

Descartes may not have known the particular works mentioned here, but he was certainly conversant with Archimedes and Archimedean geometry. Moreover, the postulates mentioned in the previous passage elaborate upon those inserted by Archimedes at the beginning of his treatise *On the Sphere and Cylinder*, which Descartes certainly knew.[55] Thus, it seems plausible that Descartes did not merely buy Aristotle's claim on the absolute impossibility of comparing straight and curvilinear segments, but only denied the possibility of solving rectification problems "exactly," that is to say, by the intersection of algebraic curves (Cf. Mancosu 2007, pp. 118–119).

Descartes had certainly accepted the idea that arclengths could be measured using infinite polygonal approximations, as in the Archimedean treatise, and that the circle could be squared, even if only using mechanical procedures such as the bending of a string around a circle, or more sophisticated but still ungeometrical methods having recourse to special curves, such as the quadratrix or the Archimedean spiral, for instance. Descartes' knowledge of these curves mainly depended on the presentations given by Pappus and Archimedes, who accounted for them in terms of the composition of two simultaneous motions, uniform in time. Because of this kinematic element entering their definitions, neither the quadratrix nor the spiral could fit the bill of geometricity, but rather fell into mechanics, together with the solutions offered thereby.[56] Moreover, as it is easy to ascertain, the characterizing properties of these curves or "symptoms," in the terminology of Greek mathematics, involved equalities between segments and arcs. But, as we know, Descartes held that no ratio between straight and curved lines was expressible by finite polynomial equations. Therefore, mechanical curves could not in any way find a place in the algebraic universe of Cartesian geometry.

1.7 Are All Rectifications Algebraically Impossible?

Descartes' belief about the impossibility of finding an algebraic rectification of a curve is true concerning the circle and the other conic sections, but false in general. Counterexamples were found about two decades after the publication of the *Géométrie*.

[54]Clagett (1964, pp. 170ff.).

[55]Dijksterhuis (1987, p. 145).

[56]Descartes (1897–1913, Vol. 6, p. 390).

Around 1657, two mathematical students, William Neile (1637–1670) and Hendrick van Heuraet (1634–1660?), discovered that the arc-length of an arc of the curve with equation $y^2 = \frac{x^3}{a}$, also known as a "semicubical parabola," can be expressed algebraically.

This result can be easily verified using calculus. Setting, for simplicity's sake, $a = 1$ and $\frac{dy}{dx} = \frac{3}{2}x^{\frac{1}{2}}$, we can express its arc-length elements as $ds^2 = (1 + \frac{9}{4}x)dx^2$, and the length of an arc s as $\frac{1}{2}\int \sqrt{4+9x}dx$. The resulting integral can be evaluated by elementary algebra. However, the discovery of the first algebraic rectifications of algebraic curves was by no means a trivial fact in Seventeenth Century mathematics, and turned out to be at the centre of an important dispute, one mainly fuelled by John Wallis and Christiaan Huygens. Incidentally, these are two main characters who will also take part in the debate on the validity of several propositions of Gregory's *Vera circuli et hyperbolae quadratura*, which shall be discussed in the next chapter. However, whereas, in Gregory's affair, Huygens and Wallis ended up being allies, or at least shared similar views about said affair, in the dispute about rectifications, they fiercely stood on opposite sides. The role played by Wallis was also different in the two controversies. While, in the case of Gregory, he was summoned by the Royal Society as a judge, in the rectifications affair, he ignited the priority dispute and played an active part in defending the cause of British mathematicians against Huygens' claims. Since the fascinating story of this controversy will enter our discussion only marginally, I shall not examine it in detail, but address the reader to Yoder (1988) and Beeley (2008) for a more careful and colourful account.

As far as the story goes, in August 1657, Wallis received a letter from the secretary of the Royal Society, William Brouncker, containing a theorem on the rectification of the semicubical parabola, which can be easily converted into a constructive procedure to find the length of any arc of the said curve (Fig. 1.9). The proof of the theorem was credited to Brouncker himself, who had used Cartesian algebra, but the discovery was promptly credited to William Neile, at the time, a student at Wadham College. Wallis took the care to publish both Brouncker's and Neile's geometrical proofs in his treatise *Tractatus duo. Prior de cycloide et corporibus inde genitis. Posterior epistolaris; in qua agitur de cissoide, et corporibus inde genitis* (1659). These results also appeared in a letter addressed to Huygens, who had known, in the meantime, about another rectification of the same curve. In 1659, in fact, Hendrick Van Heuraet had published, in the second Latin edition of Descartes' *Géométrie*, a "Letter on the transmutation of curves into straight lines" (*Epistola de transmutatione curvarum linearum in rectas*, dated January 13, 1659), in which he too successfully rectified an arc of a semicubical parabola.[57] No extant document indicates that Van Heuraet had any previous knowledge of Neile's, Brouncker's

[57]Descartes (1659, Vol. 1, pp. 517–520). For English and Dutch translations of van Heuraet's piece, see: Grootendorst and Van Maanen (1982). For an attempt at the reconstruction of van Heuraet's life and his mathematical achievements, see Van Maanen (1984) and Yoder (1988, Chapter 7).

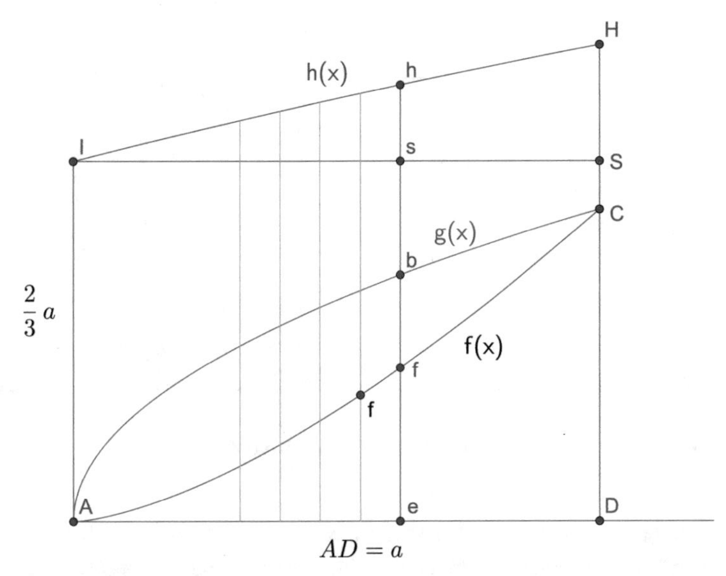

Fig. 1.9 Neile's rectification. In the figure, let $AD = a$, $DS = \frac{2}{3}a$ and $\alpha = 1$ be given. A parabola $g(x) = x^{\frac{1}{2}}$ is also given, and a second curve $f(x) = \frac{x^{\frac{3}{2}}}{a}$ is constructed, so as to satisfy the equation $\frac{2}{3}af(x) = \int g(x)dx$. In other words, the associated curve f can be traced pointwise, by dividing the parabola into thin rectangular strips, and constructing, for each parabolic section eb, the corresponding ordinates $f(x)$. In the figure, there is also a second, auxiliary parabola h of equation $y = \sqrt{\frac{4}{9}a^2 + g(x)^2}$. Neile proved that the rectangle formed by the acrlength $\widehat{AC}$ and the segment AD is equal to the area $ADHI$. Since h is a parabola, the length of AC can be computed using Cartesian algebra. Neile's purely geometrical proof was indeed rewritten by Brouncker using algebra (Wallis 1659, pp. 92–93. See also Lehay 2016)

or Wallis' work, so that Neile's and Van Heuraet's rectifications should be considered a case of independent discoveries.

The heuristic intuition in the backbone of Neile's and van Heuraet's discovery of rectifications consists in conceiving a curved line as a polygon with infinitely many sides. In this way, the length of the whole arc can be computed by adding up the small polygonal sides that form the curve. The theoretical basis of this rectification procedure was offered by Cavalieri's method of indivisibles, possibly the most successful technique available for solving integration problems before the advent of calculus. Cavalieri considered the surface of a plane figure or region (or the volume of a solid body) to be the aggregate of all of the chords (respectively, all of the planes) intercepted within the bounds of the figure when we trace the infinitely many parallel lines that cross the figure itself. On the grounds of this suggestive interpretation, Cavalieri could state a principle that still bears his name: if two plane surfaces are cut by a system of parallel lines, which intercept corresponding equal chords over each figure, then their surfaces are also equal. If corresponding chords have a constant ratio, then the surfaces too entertain the same ratio. To avoid paradoxical

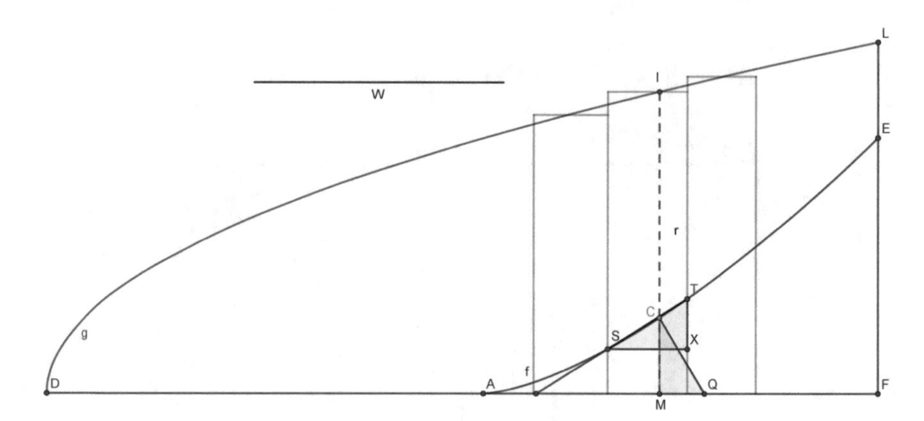

Fig. 1.10 Van Heuraet's rectification of the semicubical parabola

violations of homogeneity, which may surge when a two-dimensional figure is conceived as composed by one-dimensional objects, it became common, especially in the British mathematical tradition, with Wallis, to re-interpret Cavalieri's method and regard a surface not as the aggregate of all of its ordinates without breadth, but rather as the sum of infinitely many rectangles with infinitesimal width.

In his letter, Van Heuraet follows a similar procedure to prove a general theorem about rectifications, from which he derives the solution for the specific rectification of the semicubical parabola.[58] Let two curves be given, namely, f and g (Fig. 1.10), such that, for any arbitrary point C on the first curve, there exists a point I on the second curve, which satisfies the following proportion: $MI : W = CQ : CM$ (W is a segment with constant length, and CQ the normal to f at C). Then, the area of the curvilinear figure $GHIKL$ is equal to the rectangle contained by the segment W and by a segment equal to the length of the arc $\widehat{AE}$. Van Heuraet's proof depends on the similarity between the right-angle triangle CMQ, formed by the normal CQ to the first curve at a point C, the ordinate CM and the segment on the axis MQ, and the infinitesimal right-angle triangle SXT, whose hypothenuse ST is tangent to the curve at C. As in the case of Neile's rectification, Van Heuraet conceives the arc of the curve between A and E as made up by patching the infinitesimal segments ds together, and the corresponding curvilinear figure $GHIKL$ as composed by thin rectangular strips whose sides are, respectively, the variable segment MI and the constant segment dx; it follows at once that the rectangle bounded by W and by a segment equal to the arc-length of $ABCDE$ is equal to the area $GHIKL$. If we now apply this theorem to the algebraic curve of equation $y^2 = \frac{x^3}{a}$, i.e., to a semicubical parabola, then its companion curve is a parabola (as in Fig. 1.10). Therefore, the arc AE of the semicubical parabola also proves to be rectifiable using Cartesian geometry.

[58] Panza (2005, pp. 119–132) and Yoder (1988, p. 125, p. 126).

A second, important result obtained by Van Heuraet regards the equivalence between the rectification of an arc of a parabola and the quadrature of an hyperbolic sector. This result can be easily verified, and Van Heuraet does not make any pains to prove it, stating it as a simple corollary. Since the algebraic quadrature of the hyperbola is an impossible problem, as Leibniz would prove, another geometrical impossibility stems from Van Heuraet's corollary: the rectification of the parabola by algebraic means.

Van Heuraet's discovery was hailed by the Cartesian mathematicians of the 1660s and early 1670s, and by Huygens *in primis*, as one of the ripest fruits of the Cartesian method. One might object that neither Neile's nor Van Heureat's procedures really complied with Descartes' requirements, specifically with the restriction mandating the exclusive use of finitary objects in geometrical proofs. However, as the above examples clearly show, reasoning with infinitesimals enters Neile's rectification or Van Heuraet's *Epistola* only in regard to the analysis of the problem, while no infinitary objects appear in their solution to the rectification of the semicubical parabola, which is expressed by ordinary Cartesian algebra. We encounter a similar situation in the case of the quadrature of the parabola. As an alternative route to the classical method of proof based on the double *reductio ad absurdum*, it was customary in the Seventeenth Century to solve the quadrature of the parabola using indivisible, infinitesimal, or infinite series. In all of these cases, infinitary objects were used only as auxiliary tools, leading to a solution of the problem in more straightfoward ways than use of the rigorous proof technique based on a double reductio ad absurdum would have. Nevertheless, in the synthesis or construction of that problem, no infinitary objects were required: as we know, the area of a parabolic segment can be constructed, with exactness, using only the ruler and the compass.

Van Heuraet thus exceeded the expectations of Descartes himself and undermined the latter's belief that curved lines could not be compared with straight ones. Huygens underlined precisely this point when he presented Van Heuraet's discovery in July 1659:

> If you have not yet looked upon the new edition of Descartes' Geometry prepared by Schooten, I have to tell you about the discovery of an outstanding young man of ours whose name is Heuraet ... indeed after further investigation he found out a kind of curves (among those judged to be geometrical), and taught that straight segments could be made absolutely equal to them ... Thus you must not think any more that to find a curve equal to a straight line is against nature.[59]

On the other hand, references to Descartes' opinion on the impossibility of rectifying curves are seemingly absent from Wallis' concerns, as it appears from his account of

[59]Huygens (1888–1950, vol. 2, p. 436): "Quod si autem novam editionem Geometriae Cartesij quam Schotenius procuravit nondum vidisti est quod tecum communicem insigne inventum juvenis cujusdam nostratis Heuratij nomine ... deinde vero ulterius inquirens curvarum quoddam genus reperit (et quidem earum quae Geometricae censentur) quibus rectas lineas absolute aequales constitui posse docuit ... Non est itaque quod ultra naturae repugnare existimes curvam rectae aequalem inveniri".

rectifications and his exchanges with Britsh mathematicians and fellows from abroad, such as Huygens. On the contrary, Huygens was keen on stressing how Van Heuraet "has shown that straight lines can be absolutely equal to other curves, from the kind of those we receive in Geometry,"[60] by which he certainly meant the geometry of Descartes. A similar point was stressed, several years later, in Huygens' work on pendulum clock, the *Horologium Oscillatorium*. Returning to the priority dispute between Neile and Van Heuraet, Huygens precisely remarked how the former had failed to grasp the nature of the curve that he was going to rectify.[61] Huygens could not mean that Neile ignored that the curve was a semicubical parabola. More likely, Huygens meant that Neile had failed to understand the geometrical nature of this curve in the background of Descartes' canon of geometricity. What seems a mere passing remark, in fact, hides a deeper criticism of Neile, Wallis and Brouncker together: even if these mathematicians might have been the first to work out the rectifications of certain algebraic curves, they did not place any emphasis on the significance of this result in the background of Cartesian geometry. On the contrary, a mathematician such as Huygens probably saw, in the rectification of algebraic curves, not only an important first-rate mathematical discovery, but a discovery whose implications were far-reaching, as they undermined Descartes' strongly-worded opinion about the impossibility of solving the rectification problem exactly, i.e., algebraically.

The indifference to Descartes' opinion on the impossibility of rectifications shown by Neile and his colleagues appears to be an instance of a more general attitude common among early-modern British mathematics, in opposition to the continental one, towards Descartes' methodological or metamathematical views in the *Géométrie*. In a similar way, for instance, James Gregory downplays another cornerstore of Descartes' foundations of geometry: the distinction between geometrical and mechanical curves.

But in what sense, precisely, was Van Heuraet's result perceived as so revolutionary with respect to Descartes' geometry, at least by the mathematicians on the continent? By refuting Descartes' beliefs, the algebraic rectification of certain algebraic curves undermined the bounds of geometry as they were set forth in the *Géométrie*, and, for this very reason, the rectification in question opened up the possibility that problems excluded from Cartesian geometry could be solved by exact means. This would be the case of the circle-squaring problem. Even if one was obviously not entitled to infer from the algebraic rectification of the semi-cubical parabola the possibility of rectifying the circumference of the circle, the former result could have legitimated the search for the latter, which now did not appear as absurd as it was to Descartes' eyes. As we shall see, geometers of renown like Christiaan Huygens even believed that the ratio between the circumference and the diameter might perhaps be expressed by a rational or an irrational, surd number.

[60] "Rectas aliis curvis absolute aequales ostendit, ex earum genere quas in Geometriam recipimus" (Wallis 2003, p. 582).

[61] Beeley (2008, p. 294).

Within this historical context, the solvability of the quadrature of the circle and the determination of the exact boundaries of Descartes' method was the task that Gregory, and later Leibniz, tried to decide mathematically. This effort, which eventually led to a negative answer, will constitute the theme of the next two chapters.

2 James Gregory and the Impossibility of Squaring the Central Conic Sections

2.1 A Seventeenth Century Controversy on the Impossibility of Squaring the Circle

The *Vera Circuli et Hyperbolae Quadratura in sua propria proportionis specie inventa* (Gregory 1667, hereinafter *VCHQ*) was James Gregory's debut work in the domain of quadrature problems. It was published in Padua in 1667 and reprinted a few months later, in the spring of 1668, as an appendix to another treatise, the *Geometriae Pars Universalis* (Gregory 1668, hereinafter *GPU*).[1]

In the *VCHQ*, Gregory allegedly proved the impossibility of squaring a sector of the circle, the ellipse, and the hyperbola (here illustrated in Fig. 2.1) using finite, polynomial algebra, and presented a novel method for computing their areas by an infinite recursive procedure.

The first edition of the *VCHQ*, printed in 150 copies, circulated among Gregory's acquaintances, distinguished mathematicians and learned societies. Gregory also sent a copy of his work, together with an accompanying letter, to Christiaan Huygens, a renowned expert in the study of the circle-squaring problem. Gregory was eager to know Huygens' opinion, as his letter shows:

> I send you once more another small fruit of my intellect, which I obstinately ask you to receive and consider with your very perfect and acute judgment, and I ask you to let me know your opinion, which I await above all and I ask you vehemently. In fact I know how admirably and

[1]For a full account of Gregory's mathematical works, see Malet (1989a). See also Turnbull (1939, p. 45) and Huygens (1888–1950, Vol. 6, p. 154). The book can be also found in Huygens (1724).

D. Crippa, *The Impossibility of Squaring the Circle in the 17th Century*,
Frontiers in the History of Science, https://doi.org/10.1007/978-3-030-01638-8_2

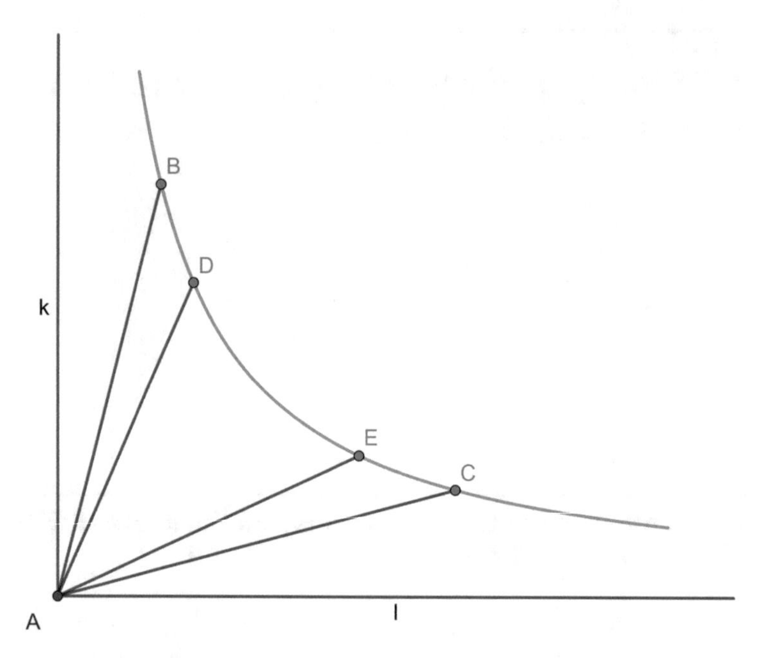

Fig. 2.1 Gregory's definition of sector applies to all conic sections with a geometrical centre: "if in a circle, in an ellipse or in a hyperbola, we trace two segments from the centre to the perimeter of the section, we call the plane included by these segments, and by a segment of the perimeter, a 'sector' " (*VCHQ*, p. 9). For the sake of simplicity and generality, Gregory always considers sectors in which the segments are equal, such as the sectors ABC and ADE of the hyperbola with center A and asymptotes k and l, depicted in the figure above

> ingeniously you have refuted the great and elegant *Opus* of Gregory of St. Vincent on this very same topic, and how acutely you have yourself written on the same subject.[2]

In this passage, Gregory refers to Huygens' refutation of an alleged solution to the circle-squaring problem that appeared in Grégoire of St. Vincent's *Opus Geometricum*. This refutation solidified Huygens' reputation as an expert on the problem of circle-squaring.[3] Huygens also wrote a well-known work on the approximate calculation of the circumference, which appeared in 1654, *De circuli magnitudine inventa*, but Gregory was probably not familiar with it until 1668.[4]

[2]"Mitto denuo alium ingenioli mei partum, quem obnixe peto, suscipias et purissimo, perfectissimoque judicio tuo ponderes, mihique censuram tuam remittas, quam imprimis exspecto et vehementer a te peto; novi enim ego quam pulchre et ingeniose Gregorii a Sancto Vincentio magnum molle opus in hac eadem materia refutaveris, et quam acute de hoc eodem argumento tu ipse scripseris." (Huygens 1888–1950, Vol. 6, p. 154). As hinted in the letter, Gregory had already sent one of his works on optics, (*De optica promota*), to Huygens.

[3]Huygens (1888–1950, Vol. XI, p. 273).

[4]See Huygens (1888–1950, Vol. XII, p. 93).

Contrary to Gregory's expectations, Huygens never dignified his younger colleague with a reply. During the spring of 1668, he instead assembled a report on the book for a meeting of the *Académie des Sciences*. The report, which appeared in print in July 1668 as a letter to the *Journal des Sçavans*, emphasized major flaws in Gregory's work, flows that substantially demolished all of the original contributions brought by the *VCHQ*.[5] This is particularly surprising, since Gregory's work had so far received favourable reviews from the members of the Royal Society. For example, in the *Philosophical Transactions of the Royal Society*, the *VCHQ* was even judged: "very ingeniously and very Mathematically written and well worthy the study of men addicted to that Science."[6] Gregory's subsequent reaction ignited a public controversy with Huygens that lasted for several months, and was mainly consigned to letters, either sent to private recipients or published in the *Philosophical Transactions of the Royal Society* or in the *Journal des Sçavans*.[7]

This controversy proceeded on two intertwined levels. On one level, Huygens and Gregory debated over the central mathematical issue at stake, namely, the impossibility of the algebraic quadrature of the circle and of the other central conic sections. On a second level, the two mathematicians crossed swords after certain critiques advanced by Huygens that Gregory understood as charges of plagiarism and a lack of originality. Although this level is less relevant to the theme of our book, it loomed large over the whole controversy and shaped it in a substantial way, to the point of sanctioning its end.

Huygens' critical observations are summarized in the letter appeared on July issue of the *Journal des Sçavans*. Huygens' first criticism regards Gregory's impossibility result, and will be examined later on, in Sect. 2.5. As for the remaining objections addressed by Huygens, the editor explained, in an introductory note to the letter:

> Regarding the method proposed by the author in order to approximate by numbers the area of the circle, Huygens said that he thought he had given given something more precise in the book titled *De Circuli magnitudine*, which he had printed in the year 1654. He added that what is said in this book regarding the area of the hyperbola and its relation with logarithms is very good, but that the men of the assembly [namely, during the session at the *Académie des sciences*] did not find it new, as they could remember that Huygens had proposed the same thing to them, and that the rule for finding logarithms has long been recorded in their Book. He also did not think that this would have appeared as a novelty to the members of the Royal Society.[8]

[5]Cf. Journal (1668), also reprinted in Huygens (1888–1950, Vol 6, p. 228; Vol. 20, p. 259).

[6]Transactions (1668, p. 641).

[7]For the public letters, see Gregory (1668a), Gregory (1668c), and Huygens (1668). The pieces of the controversy, consisting in drafts, public and private letters, are reproduced in Huygens (1724), and in Huygens (1888–1950, Vol 6., Vol. 20). See also Dijksterhuis (1939, p. 485) and Feingold (1996).

[8]Huygens (1888–1950, vol. 6, p. 231): "Pour ce qui est de la methode que l'Auteur propose d'approcher par nombres de la dimension du cercle, M. Huygens dit qu'il croyoit avoir donné quelque chose de plus précis, dans le Livre intitulé *De Circuli magnitudine*, qu'il a fait imprimer dés l'an 1654. Il ajôuta que ce qui est dit dans ce Livre touchant la dimension de l'Hyperbole & le rapport

Although Huygens had actually made no direct accusations of plagiarism, nevertheless, Gregory deeply resented these commentaries. He replied to them mainly in his book *Exercitationes Geometricae*, printed towards the beginning of the autumn of 1668. In this small book, Gregory dismissed Huygens' allegations as mere: "critiques of little weight," to which he could not reply immediately, having no knowledge of Huygens' own writings on the subject.[9]

As per his own account, Gregory could not have mentioned Huygens' work on the approximation of the circumference because he utterly ignored it when the *VCHQ* appeared. Gregory also added that Huygens' claim regarding the quadrature of the hyperbola was not backed by sufficient evidence:

> I certainly would have expected more ingenuity from such a man. Nor have I been able so far to find out whether there is anyone from the Royal Society who has recorded what Huygens had once recounted. Not only these reasons but other ones as well almost persuade me that the quadrature of the hyperbola was not known to Huygens, at least so many years ago. Indeed, mathematicians do not desire anything more in the whole geometry, and particularly anything more apt to the human usefulness, that a geometer would thus not conceal, for an interval of seven or eight years and in such a prophetic way, an investigation not only very well-known but also of the greatest usefulness.[10]

Gregory's overall strategy becomes clear through the passage above: in questions of scientific precedence, the first who publishes a discovery settles the dispute. Gregory had a public evidence-based criterion of legitimacy in mind. That is, legitimacy only came in the form of a printed or written claim that could be objectively and publicly checked, whereas personal recollections or oral declarations were subject to the whims of time and memory.

This idea was common among members of the Royal Society, who abhorred plagiarism and viewed it as one of the most shameful and gravest charges that could be made against

qu'elle a avec les Logarithmes, est fort bon; mais que Messieurs de l'Assemblée ne le trouveroient pas nouveau, puisqu'ils pourroient se souvenir qu'il leur a desja proposé la même chose, & que la regle qu'il a donné pour trouver les Logarithmes est inseré il y a long temps dans leur Registre: Qu'il ne croyoit pas non plus que cela parust nouveau à Messieurs de la Societé Royale d'Angleterre."

[9]Gregory (1668b), *Appendicula ad veram circuli et hyperbolae quadraturam*, p. 1: "Duae autem fuerunt parvi momenti, quibus tunc temporis respondere non potui; utpote in ejus scriptis nequaquam versatus." See also Huygens (1888–1950, vol. 6, p. 315).

[10]Gregory (1668b), *Appendicula ad veram circuli et hyperbolae quadraturam*, pp. 1–2: "Expectabam certe a tanto viro majorem ingenuitatem; Nec enim potui hactenus edoceri, esse quenquam e Societate Regia, qui tale quid ab *Hugenio* unquam prolatum recordatur. Non solum praedictae rationes sed etiam aliae mihi quasi persuadent Hyperbolae quadraturam, saltem ante tot annos, *Hugenio* non innotuisse; nam nihil in tota Geometria a Mathematicis adeo est desideratum, immo nec usui humano magis accommodatum; quis ergo Geometra speculationem non solum celeberrimam sed etiam maximae utilitatis 7 vel 8 annorum spatio adeo superstitiose celaret." For the quadrature of the hyperbola, see the next chapter.

someone. Huygens likely shared a similar opinion on such issues. In his replies, which circulated in the form of both public and private letters during autumn of 1668, showed utter disconcert at Gregory's reaction, and denied that his own intentions were to mount a case of plagiarism.[11]

Even after such clarifications, a few members of the Royal Society took Gregory's position quite seriously, worried by the tones that the controversy was taking. For instance, Robert Moray (1608–1673), the first president of the Society, who kept a correspondence with both Huygens and Gregory, tried to soften the dispute, arguing that Gregory's resentment was partly justified, as no allegation could do more harm to a gentleman than plagiarism. In addition to Moray, John Collins (1625–1683), an outstanding member of the Royal Society due to his vast network of correspondents (among whom was Gregory himself), showed a sympathy for Gregory and made an effort to reconcile him with Huygens. Thus, Collins argued that Gregory's and Huygens' achievements were pristine cases of independent discoveries, which should not be a cause of dispute, since: "the variety of methods does much advance inventions."[12] Collins held that no two genuinely independent discoveries of the same theorem or solutions to the same problem are exactly alike. Their difference, which could be methodological, may fruitfully open up new lines of research. Hence, not only had Gregory not copied Huygens, since he ignored the latter's work when he was composing his own,[13] but their independent discoveries were both to be hailed as important contributions to the advancement of learning.

Collins' ultimate judgment on the controversy between Gregory and Huygens did not spare the conduct of either character from criticism:

> Upon ye whole, Monsieur Hugens seems blameable for beginning these comparisons, quasi ex animo vilipendendi, as appears from his reason rendred, why Gregory's quadrature of ye Hyperbola should not seem new to ye Royal Society; on the other side it were to be wisht, that Mister Gregory had been more mild with yt generous person, who hath deserv'd well of ye republick of Learning.[14]

---

[11]Huygens to Wallis, 13 [03] November 1668: "Credebam equidem prima illa discussione mea nihil eum offensum iri, namque et non sine laude de summa operis locutus sum, et quae parum evidenter demonstrata erant, examinare concessum putabam ... ferocior factus neque exspectans quid ei reposíturus essem, acerbissimo scripto in me nihil tale metuentem homo invehitur, plagijque sese accusatum praetexens publice me mendacij insimulat." (Huygens 1888–1950, Vol. 6, pp. 280–281).

[12]Huygens (1888–1950, Vol. 6. p. 372).

[13]This is stated by Gregory in the *Exercitationes*, but also confirmed by Collins' account: "upon diligent search there hath not been any of Hugenius ... books to be found in any Stationers shop in London for 12. years past, and probably they are as scarce in Italy; which perchance might induce Mister Gregory to write a Treatise of this kind." (In Huygens 1888–1950, Vol. 6, p. 373).

[14]Huygens (1888–1950, Vol. 6, p. 376).

Ultimately, it seems that the "republic of Learning" was more concerned by the harshness of the exchanges between the two mathematicians than the content of the controversy, as little interest was given to the diverging mathematical opinions between Gregory and Huygens. Disputes such as theirs were not unusual during the early-modern period.[15] For this reason, many members of the Society had decided to adopt a conciliatory attitude, seeking to steer clear of any quarrel in a matter of science, and they applied this attitude towards Gregory's and Huygens' heated correspondence. To aggravate the situation, neither Gregory nor Huygens had managed to keep their disagreement within certain tacit bounds of decency, and had resorted to foul play, such as innuendos of plagiarism, name-calling, and unproven claims of precedence.

These were, I think, the main reasons why both Gregory and Huygens were forced to close the controversy. By February 1669, Henry Oldenburg, secretary of the Royal Society, refused to host more contributions on the topic in the *Philosophical Transactions of the Royal Society*.[16] This decision brought the public exchanges between Gregory and Huygens to an end without a proper resolution having been reached.

As a result of the decision to end the correspondence, the question about the correctness of Gregory's claim of the impossibility of squaring the central conic sections remained open. Huygens both denied the validity of the impossibility proofs given by his opponent and expressed doubts concerning the impossibility of squaring the circle, while Gregory insisted on defending his position, even if his arguments were not always mathematically cogent, and sometimes were themselves tainted by bitter, polemic tones. The following response to Huygens exemplifies such polemics:

> We do not lack those who ... haughtily reject the very subtlest parts of a discovery, not without damaging its glory, and in the meantime despise the answers, believing to be bright, and infallible among the ignorants, as much as they feel at ease among the learned ones ... Men are so pleased with their own authority, that they believe that nothing new (which is of some importance) can be found by new authors.[17]

In order to clarify the matter at stake, in the following sections, I shall discuss in detail Gregory's method of quadratures and his "subtlest" discovery, namely, his impossibility theorem (Sects. 2.2, 2.3 and 2.4). I shall then examine Huygens' main objections to the validity of Gregory's impossibility theorem (Sect. 2.5) and, in the conclusion, I shall

[15] A survey of other disputes can be found in Beeley (2008), which deals with John Wallis and Huygens, or Jesseph (1999) on the controversy between Hobbes and Wallis on the quadrature of the circle.

[16] See also below, Sect. 2.5.3.

[17] Gregory (1668b), *Praefatio*, unnumbered sheets: "Imo non desunt qui ... reliqua longe subtiliora non sine famae suae iactura fastuose reijciunt; responsa interim spernunt, praeclarius existimantes, infallibiles haberi apud ignaros, quam ingenui apud doctos ... Nam adeo arridet hominibus authoritas, ut nihil novi (quod alicujus momenti sit) a novis authoribus inveniri credant."

propose some considerations on Gregory's views on impossibility in mathematics in general, as well as specifically in relation to geometrical exactness (Sect. 2.6).

2.2 Gregory's "Second Kind of Analysis"

In the introductory remarks to Huygens' first critical report from July 1668, the author, probably Huygens himself or Jean Gallois, director of the *Journal des sçavans*, emphasizes the novelty of Gregory's approach to the circle-squaring problem with these words:

> Since the problem of the quadrature of the circle has always been famous because of its difficulty, the most acute geometers of all times have committed themselves to find the solution ... but even if they came so close to it to the point that, in squaring a circle as large as the whole Earth, we are assured that the error will be smaller than a hair, no one has achieved the ultimate accuracy demanded by geometry. The author of this Book deals with this topic [i.e., the quadrature of the circle and the hyperbola] in a new way. He firstly undertakes to show that the ratio of the Circle to the square of the diameter was not of the kind he names "analytical"; it is in vain that we try to express it in terms received in Geometry.[18]

Huygens and his fellow mathematicians thus clearly saw that the most remarkable result of Gregory's work was the impossibility of expressing the ratio between the circle to the square of the diameter in "received terms," namely, the five arithmetic operations of addition, subtraction, multiplication division and extraction of roots, applied to known or given quantities.[19]

Gregory's work was not limited to the circle-squaring problem, understood in the classical sense of constructing a polygon equal to a given circle. As I mentioned in the introduction, problems of quadrature were explicitly distinguished, from the second half of the Seventeenth Century, in two types of mathematical questions: the "definite"/"particular" and the "indefinite"/"universal" quadratures. A "definite" quadrature denotes the problem of finding the total area enclosed by a given curve, or the area of a pre-assigned portion of it. In modern terms, this is equivalent to computing a circle-measuring

[18]Journal (1668, p. 76): "Comme le problème de la Quadrature du cercle a toujours été si celèbre à cause de sa difficulté, les plus subtils géomètres se sont de tout temps appliqués à en chercher la solution ... Mais quoy qu'ils en ayent approché si près, qu'on est assuré qu'on ne se tromperait pas seulement de l'epaisseur d'un cheveu dans la Quadrature d'un Cercle aussi grand que toute la Terre; neantmoins personne n'a pû encore arriver à la dernière precision que demande la Géométrie. L'Auteur de ce Livre traitte ce sujet d'une manière nouvelle. Il entreprend premierement de faire voir que la raison du Cercle au quarré du diametre n'estoit pas ce qu 'il appelle analytique, c'est en vain qu'on tâche de l'expliquer en des termes reçeus dans la Geometrie."

[19]This judgment was shared by British mathematicians too. As we read, for instance, in a letter from John Collins to John Pell dated November 1668: "the grand Designe of the booke [namely, the *Vera circuli et hyperbolae quadratura*]" was to prove the impossibility of squaring the central conic sections "by any Analyticall Operations whatsoever" (Wallis 2012, p. 7).

integral between given bounds. The classical problem of squaring the circle, which requires measuring the area of the whole circle or constructing a square of equal area, is actually an example of a definite quadrature. The "indefinite" or "universal" quadrature was the problem of finding the area of a portion of a curve cut by a line arbitrarily traced.[20] In modern terminology, the indefinite quadrature of the circle is equivalent to finding the antiderivative of a circle-measuring integral. Although the translation into the language of calculus clarifies the matter for us, we must also bear in mind that early-modern mathematicians during the 1660s did not have a precise understanding of the distinction between the problems of the indefinite and the definite quadratures of the central conic sections. As I shall discuss later on, this point is quite important in order to seize the roots of the disagreement between Huygens and Gregory about the content of Gregory's alleged impossibility theorem. Moreover, one of the outcomes of the controversy between the two mathematicians, to which Leibniz also greatly contributed, was a better understanding of the nature of quadrature problems and their mutual independence.

In the *VCHQ*, Gregory focused on the problem of the indefinite quadrature, not of the circle alone, but of the central conic sections as well. The *VCHQ* is organised in a traditional manner: after the preface, written in the form of a dedicatory letter to the "friendly reader," Gregory listed ten definitions and two postulates (*petitiones*), followed by the thirty-five propositions (divided into theorems, problems and *scholia*) that comprise the whole treatise.

My survey of the *VCHQ* will be limited to the first eleven propositions (plus the definitions and postulates), which occupy less than half of the whole treatise. However, they can be seen as forming a unitary body. On one hand, they are somehow preparatory to Gregorys' impossibility argument; on the other hand, the saliency of these propositions is recognized by Gregory himself, who admitted their theoretical import in contrast to the remaining ones, which were merely added "for facilitating the practice of his new techniques."[21]

In the "friendly reader" preface to the *VCHQ*, Gregory expounded the motivations and rationale of his work:

> I have wondered sometimes, my friendly reader, whether Analysis with its five operations were a sufficient and general method for investigating all proportions between quantities, as Descartes seemed to affirm in the beginning of his geometry; if it were so, it would be possible, by its aid, to exhibit the so illustrious quadrature of the circle: thinking with this idea in mind, I could easily perceive, from the properties of the circle so far discovered, that no analysis

[20]Cf. Lützen (2014, p. 217).

[21]This point was made, as a clarifying remark, by Gregory to Collins, on March 26, 1668. See Turnbull (1939, p. 51). With the expression "facilitating the practice," Gregory might have been referring to the new procedures for the approximate measurement of the area of the circle and the hyperbola and for the calculation of logarithms, exposed in his treatise, especially in propositions XXIX–XXXIV.

> could be construed so as to serve such a structure [namely, the structure of the problem]: then there came to my mind a second kind of analysis, while I was searching for other ones (the first, concerning the circle, was indeed known to the laymen). Through these I understood the convergent series of polygons, whose termination [*terminatio*] is a sector of the circle, where, immediately, I saw some trace of analysis. Thereafter, having considered the natures of the convergent series not only in the easier cases, but also in general, and having reduced the properties predicted for the circle to the ellipsis and the hyperbola with little trouble, the quadrature of every conic sections seemed infallible to me.[22]

As appears from the first lines of the above excerpt, Gregory started his book with a methodological question: was Descartes really justified in claiming to possess a method that could solve all of the problems of geometry? Gregory was probably thinking of the opening statements of the *Géométrie*, in which Descartes gave the outline of a method by which, he declared, all problems of geometry could be solved by reducing them to problems about line segments, and then to algebraic equations. Taking Descartes' claim literally, Gregory took the problem of squaring the circle as a test-case in order to probe the generality of the Cartesian method of analysis. This choice is reasonable, since, as I have argued in the introduction, Descartes' belief that straight lines and curvilinear arcs could not be compared had lost much of its credibility by the end of the 1660s. Therefore, at the time, the question as to whether Descartes' method could encompass the rectifications of all algebraic curves, and particularly the one of the circle, was a truly open question (see Sect. 1.7).

Thus, after having pointed out the inadequacy of the methods "known to the laymen" ("*vulgo ... cognita*") in order to deal with the circle-squaring problem, Gregory announced the discovery of a "second kind of analysis", which he considered more apt for studying the quadrature of the circle and the other conic sections as well. By the former, known methods, Gregory was possibly referring to either the classical Archimedean techniques or to the improvements they underwent from the Sixteenth and Seventeenth Centuries up to Gregory's time. On the other hand, by his "second kind of analysis", Gregory meant the general method that he would expound upon in the first ten propositions of his book and that we will investigate below.

---

[22] *VCHQ*, p. 4: "Mecum alinquando cogitabam, amice lector, num analytica cum suis quinque operationibus esset sufficiens, et generalis methodus investigandi omnes quantitatum proportiones, ut in initio suae Geometriae affirmare videtur Cartesius; si enim ita esset, possibile foret ejus ope toties decantatam circuli quadraturam exhibere: cumque hac mente revolverem, facile percepi ex hactenus repertis circuli proprietatibus nullam posse analysin institui tali structurae inservientem: deinde mihi alias quaerenti incidit in mentem huius secunda, prima enim in circulo vulgo est cognita: ex hisce percepi seriem polygonorum convergentem, cujus terminatio est circuli sector, ubi statim vidi aliquod analysios vestigium. Deinde serierum convergentium naturis non solum in facilioribus quibusdam casibus, sed etiam in genere consideratis, et praedictis circuli proprietatibus ad ellipsim et hyperbolam nullo negotio reductis, infallibilis mihi videbatur omnium sectionum conicarum quadratura."

As Gregory recounted in the preface, he was probably confident, in the first instance, that such a "new analysis" could eventually lead to an exact solution to the quadrature of the central conic sections, but was obliged to develop a different strategy after running into "unspeakable difficulties:"

> Considering that the spirit of analysis, as well as that of common algebra [*analysios esse sicut algebrae communis*] is not merely to solve problems, but also to prove their impossibility (if there is the necessity); and since I have found unspeakable difficulties in the first, I turned to the second, which I have obtained certainly beyond my expectation. In fact, I shall reveal (a task that I proposed to myself from the beginning) not only the true and legitimate quadrature of the circle in its own kind of proportion—an untouched kind of proportion, unknown before in the domain of geometry—but of all the conic sections. [*et integram proportionis speciem ante incognitam orbi Geometrico*].[23]

While Gregory recognized the novelty of his approach, he also based it on the tradition of analysis and algebra. We may wonder about the sources of Gregory's idea, stated clearly in the passage above, that an impossibility result could be an intended outcome of analysis. One hypothesis is that Gregory had in mind the very beginning of Pappus' third book of the *Collection*, the same quoted earlier in the introduction, where it is remarked that the tasks of the inquirer include that of determining: "what is possible and what is impossible, and if possible, when and how and in how many ways it is possible."[24] Gregory might also have thought of existing examples of impossibility results, like the incommensurability between the side and the diagonal of a square or its counterpart in algebra, i.e., the proof of the irrationality of $\sqrt{2}$. Several impossibility results were also commonly encountered in algebra, like the impossibility of expressing the root of certain equations as integer, fractional or surd numbers. The latter case in particular resonates with Gregory's reference to "common algebra" and with his remark according to which geometry was required to be enriched with a new kind of proportion. Negative, surd and imaginary numbers were, in fact, standard currency among Sixteenth Century algebraists, and had initially been introduced without great worries for their ontology (their nature would become a matter of concern later on, during the Seventeenth Century, as we shall see at the end of this study). Just as other mathematicians made use of these new kinds of numbers, Gregory

[23]*VCHQ*, p. 4: "Sed animo revolvens analysios officium esse sicut algebra communis, non solum problemata resolvere, sed etiam eorum impossibilitatem (si opus sit) demonstrare; eumque in primo difficultatem indicibilem expertus essem, ad secundum me converti, quod certe supra votum successit; non enim solius circuli (quam mihi ab initio proposueram) sed omnium sectionum conicarum veram et legitimam in sua proportione specie quadraturam, et integram proportionis speciem ante incognitam orbi geometrico patefacio."

[24]Cf. also Pappus' classical account of analysis and synthesis in Book 7 of the *Collection* (Pappus 1986, pp. 82–83). I would like to stress Pappus' explicit reference to the eventuality that an analysis might result in false or impossible statements (depending on whether a theorem or a problem is at stake). In such cases, a synthesis would be unnecessary, and our reasoning process would terminate when the falsity or the impossibility is reached.

required that arithmetic also be enriched with a new kind of operation: the construction of convergent series, as we shall see later in this chapter.

Because of its explicit focus on impossibility, the *VCHQ* has been viewed as a forerunner of later developments in mathematics.[25] According to common historiography, in fact, impossibility results would gain popularity only in the Nineteenth Century, when they were proved by means of algebra and through the use of techniques extraneous to the period we are considering. For this reason, the following remark from the preface of the *VCHQ* might be read as an anticipation of a trend of research that would gain popularity centuries later:

> As regards these questions, there are such vast fields of research: [which include] ... the proof that the *mesolabium* cannot be superseded by ruler and compass, that composed equations cannot always be reduced to pure ones, and in which cases they can, and which curve of the minimal kind is necessary for the mechanical solution of given equations, with countless similar cases, which are cognized as impossible by the most expert geometers on the grounds of analysis, and are daily searched for in vain by the less skilled ones.[26]

Yet, it should be stressed that Gregory's agenda concerns examples of impossibility results quite different from those to which we are accustomed in modern mathematics. His agenda first includes the impossibility of solving by ruler and compass problems solvable through the "mesolabe," namely, solid or higher problems[27]; secondly, the impossibility of factoring certain equations[28]; thirdly, the general problem of determining the curves of minimum degree necessary for the construction of a given problem. The latter problem does not directly concern impossibility, but is related to it, since a method capable of assessing the most adequate curves for any given problem ought to precisely determine which curves cannot solve a problem at hand.

[25]Cf., for instance, Dehn and Hellinger (1943, p. 160): "A modern mathematician will highly admire Gregory's daring attempt of a 'proof of impossibility' even if Gregory could not attain his aim. He will consider it a first step into a new group of mathematical questions which became extremely important in the Nineteenth Century."

[26]*VCHQ*, pp. 5–6: "cum in his tam late pateant inventionum campi ... quod mesolabium non posse perfici ope regulae et circini, item quod non semper et quando aequationes affectae possunt reduci ad puras, item quod necessaria fit ad minimum talis generis curva ad mechanicam talium aequationum resolutionem, cum talibus innumeris, quae a praestantioribus geometris impossibilia esse deprehenduntur ex analysi, et a rudioribus quotidie et frustra quaeruntur." See also Dehn and Hellinger (1939, pp. 474–475).

[27]The mesolabe is an instrument, attributed to the Greek mathematician Eratosthenes, for solving the problem of inserting two proportionals between two given segments. Descartes extended its application to a higher number of mean proportionals (Cf. Sasaki 2003, p. 112).

[28]This algebraic problem is also related to ruler-and-compass constructibility, as is clearly evident throughout Book III of the *Géométrie*, in which Descartes discussed the factoring of equations of degree 3 or higher.

Moreover, Gregory was also keen to admit that a proof of impossibility would be fully satisfactory only when reduced to geometrical terms. As we shall see, this will clearly not be the case for the proof that it is impossible to square the central conic sections analytically, for which he admitted: "it is certainly true that I have not reduced this proof to a pure geometrical expression. In fact, in order to do this, one needs a non-small volume about the reciprocal relations of analytical quantities between themselves, and, generally speaking, about incommensurability."[29]

Gregory thus imagined that a purely geometrical proof of the impossibility of the analytical quadrature of the circle ought to stem from a deeper study of quantities composed from given quantities by a finite combination of the five arithmetical operations. This *desideratum*, certainly at odds with the modern treatment of the impossibility result, which hinges on the algebraic approach, betrays the specificity of Gregory's approach and motivations, which are different from ours. In the next section, I shall review Gregory's attempts to transform the Archimedean geometrical method of approximations into "an algebraic one ... a sort of calculus," as Dehn and Hellinger nicely put it.[30] This transformation turned out to be an instrumental step in the formulation of the theorem of impossibility presented in the *VCHQ*.

2.3 Introducing Convergent Sequences

2.3.1 Polygonal Approximations

The structure of the second kind of analysis devised by Gregory can be tentatively reconstructed both from the cursory account presented in the preface and from the content of the *VCHQ* (especially, Propositions I–V, pp. 1–15).

I shall distinguish two steps constitutive of this analysis. The first one consists in the elaboration of a geometrical approximation method, detailed in the opening lines of *VCHQ*, for approximating the area of the circle by the successive construction of in- and circumscribed polygons. The second step consists in extrapolating an infinite double sequence (or "convergent series" in Gregory's terminology) from the previous geometrical process.

The geometrical starting point of Gregory's analysis is a variant of the Archimedean polygonal procedure, consisting in considering areas instead of perimeters. It is therefore

[29] *VCHQ*, p. 5: "Verum certe est me hanc demonstrationem integram ad phrasem geometricam non reduxisse, nam ut hoc perficiatur, opus est non parvo volumine de quantitatum analyticarum mutuo inter se relatione et incommensurabilitate in genere."

[30] Dehn and Hellinger (1943, p. 469).

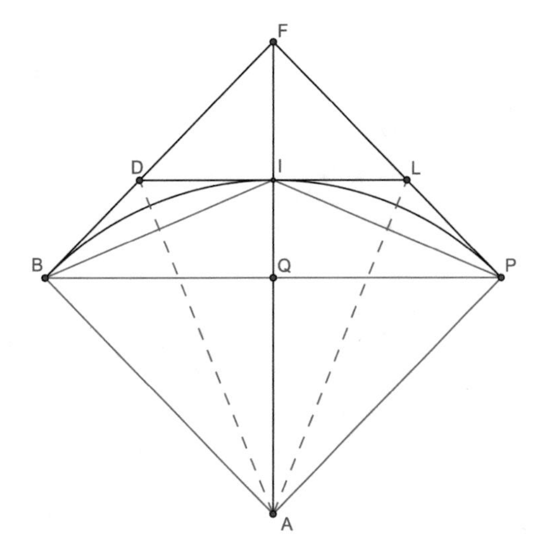

Fig. 2.2 The construction of a sequence of in- and circumscribed regular polygons to the circular sector $\widehat{APB}$

applicable to circular, elliptical and hyperbolic sectors, that is to say, to any sector of a central conic section.[31]

Gregory's construction is presented in the first propositions of the *VCHQ* (namely, *VCHQ*, pp. 11ff.) and can be summarised according to the following scheme (Figs. 2.2 and 2.3):

1. Let a conic section be given with center A. Trace the chord PB. A sector $\widehat{APB}$ will then be constructed. Trace PF and BF, both tangents to the conic, and join points F and A so as to yield point Q, an intersection between segments FA and PB, and point I, an intersection between FA and the arc delimiting the sector. In the case of a circle or an ellipse, this construction yields a triangle ABP inscribed in the sector and a trapezium $ABFP$ circumscribed to it. The same construction can be applied to a hyperbola (Fig. 2.3). In this case, though, the triangle will be called "circumscribed" ("*circumscriptum*") and the trapezium "inscribed"("*inscriptum*") to the sector.[32]
2. The same constructional procedure specified above can be applied to sectors $\widehat{BAI}$ and $\widehat{IAP}$, so as to obtain a second inscribed (or circumscribed, in the case of the hyperbola)

[31] Gregory's general method for the squaring of the central conic sections fits surprisingly well with Newton's results obtained in the 1666 treatise *De Methodis*. Newton's achievement concerned the non-algebraic quadrability of a large class of what we today call 'elliptic functions', and may stand as the first step in the modern theory of elliptic functions. However, it is out of the question, I think, that Gregory could have consulted, in 1667, Newton's treatise of 1666. Thus, the two results are independent and also very different, although consistent one with the other. Therefore, Gregory's contribution in the *VCHQ* can be taken as a parallel starting point for the theory of elliptic functions, as Dehn and Hellinger also suggest (Dehn and Hellinger 1943, p. 156).

[32] *VCHQ*, p. 9.

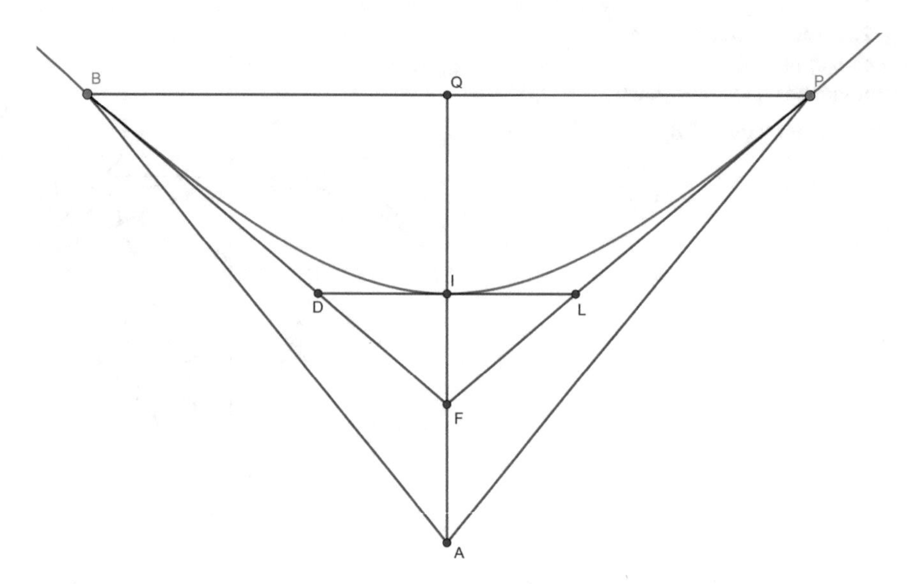

Fig. 2.3 The construction of a sequence of in- and circumscribed regular polygons to the hyperbolic sector $A\overset{\frown}{P}B$

polygon, namely, $ABIP$, and a new circumscribed (or inscribed) pentagon $ABDLP$ (or *vice versa*, in the case of the hyperbola). Moreover, if points D and L are joined with the center A, a pair of new points E and O are obtained on the perimeter of the sector, and the hexagon $ABEIOP$ can thus be traced. Similarly, by tracing a new couple of tangents in E and O, a new circumscribed heptagon can be drawn, contained in $ABDLP$, and so on.[33]

Once he had introduced this protocol for constructing two successions of in- and circumscribed polygons to a given sector, Gregory managed to represent their areas by means of a pair of infinite sequences, which can be written in modern notation as $\{I_n\}$ (the areas of inscribed polygons) and $\{C_n\}$ (the areas of circumscribed polygons). Gregory proved, furthermore, that these sequences can be recursively defined according to the following laws (rendered in modern formulation):

$$I_{n+1} = \sqrt{C_n I_n}, \tag{2.1}$$

$$C_{n+1} = \frac{2C_n I_n}{I_n + \sqrt{C_n I_n}}. \tag{2.2}$$

[33] *VCHQ*, pp. 12–13.

In this way, by reducing a geometrical recursive process to a two-term recursion, Gregory also managed to reduce the geometrical problem of squaring the sector of a conic to the algebraic problem of computing the common termination of the sequences I_n and C_n associated with the polygonal construction above. In my reconstruction, this result concludes Gregory's analysis of the problem of squaring a sector of a central conic section.

2.3.2 Analytical and Non-analytical Quantities and Operations

The backbone of Gregory's analysis is formed by the definitions given at the beginning of the *VCHQ*, and, in particular, by the definition of convergent series. In order to introduce these objects, Gregory begins by defining the notion of "composition":

> We say a quantity is composed of quantities when it derives from the addition, subtraction, multiplication, division or extraction of roots of quantities, or from any other imaginable operation.[34]

A composition is therefore an operation that, given one or several quantities, yields another quantity as a result: we may think of it as an explicit function.[35] The operations listed by Gregory in the above definition: addition, subtraction, multiplication, division or extraction of roots of quantities, are called "analytic" (df. 6) said list having probably been derived from the opening paragraphs of Descartes' *Géométrie*, where Descartes lists the four or five operations that form the core of his algebra.[36] I point out that Gregory refers generically to "quantities," without specifying whether they are magnitudes or numbers: this abstract treatment of quantities was also made possible on the grounds of Cartesian geometry, since, thanks to Descartes' definition of multiplication, the same operations could apply either to discrete or continuous quantities. From Gregory's viewpoint, moreover, Descartes' arithmetic or analytic operations did not exhaust the domain of geometry, since Gregory mentions other "imaginable" operations that might not be reducible to known ones. At this stage of the book, however, there are no examples that Gregory could cite, so that the domain of what lies beyond arithmetic operations appears to be shrouded in darkness: one task of the *VCHQ* is to explore this obscure realm and enrich the then-known geometry with new kinds of composition.

---

[34] *VCHQ*, def. 5, p. 9: "Quantitatem dicimus a quantitatibus esse compositam, cum a quantitatum additione, subductione, multiplicatione, divisione, radicum extractione, vel quacumque alia imaginabili operatione, sit alia quantitas."

[35] Youkshevitch (1976, p. 60) and Lützen (2014, p. 224).

[36] Descartes (1897–1913, Vol. 6, p. 369): "Toute l'Arithmétique n'est composée, que de quatre ou cinq opérations, qui sont l'Addition, la Soustraction, la Multiplication, la Division, et l'Extraction des racines."

On the other hand, Descartes' arithmetical operations define a domain of "mutually analytical" quantities, according to definition 7 of the *VCHQ*:

> Where quantities can be analytically composed from quantities commensurable with each other, we say that they are mutually analytical.[37]

As I will show below, this definition is crucial in order to understand Gregory's impossibility theorem. From a logical viewpoint, the predicate "being analytical" used by Gregory is a two-place symmetric relation. In other words, Gregory never speaks of a quantity that is "analytical" per se, but rather of a quantity that is analytical with respect to another quantity, or of quantities "analytical with respect to one another." In this sense, "analytical with" behaves just like "commensurable with." As per the defintion given in Book X of Euclid's *Elements*: according to Euclid, in fact, commensurability is also a two-place relation, and two magnitudes are (mutually) commensurable if they have a common measure (Book X, df. 1). For Gregory, as the above definition reports, two magnitudes are (mutually) analytical if they can be generated by applying finite sequences of arithmetical operations to given quantities, which are assumed to be mutually commensurable. For instance, if we assume a and b to be mutually commensurable magnitudes, then $c = a+b$, $d = a^3b$, $e = \sqrt[3]{ab}$ or any quantity obtained by applying finitely many sequences of arithmetical or algebraic operations will be analytic with a and b.

Against this interpretation, Scriba (1983) proposes understanding the concept of "analytical composition" in terms of a finite combination of rational operations and square-root extractions only. However, it seems to me that the notion of "analytical composition" ought to be given a less restrictive interpretation on the basis of the text itself. In fact, Gregory refers, in the df. 5 and df. 6, to "extraction of roots" (*radicum extractione*) without confining himself solely to square roots. Moreover, the restriction to quantities constructible by ruler and compass would make little sense with respect to Gregory's proposal of testing the generality of Cartesian geometry, since, as we know, Descartes included all of the magnitudes constructible by algebraic curves in geometry.

As I have suggested above, applying arithmetical operations to magnitudes, irrespective of their nature, was not a novelty by Gregory's time, as it was largely due to Cartesian mathematics. It is worth remarking, regarding this point, that Frans Van Schooten's introductory lesson in Cartesian geometry, or *Principia Matheseos Universalis* (1651), edited and published by his student Erasmus Bartholin, contains a formal exposition of the rules of Cartesian algebra or "mathesis universalis", i.e., independently from any pre-assigned reference to geometrical or arithmetical content.[38] On the other hand, Gregory may have been the first to define, in such a general and abstract way as we read in the opening pages of the *VCHQ*, a class of quantities closed with respect to arithmetical (analytical) operations.

---

[37] *VCHQ*, p. 9: "Quando quantitates a quantitatibus inter se commensurabilibus analytice componi possint, dicimus illas esse inter se analyticas."

[38] Cf. Descartes (1659, Vol. II).

This characterization of the domain of analytical quantities is achieved through the following pair of postulates, or demands ("*petitiones*"):

> 1. We demand that quantities composed analytically from given quantities analytical one with respect to the other be likewise analytical with respect to each other and also with respect to the quantities from which they were derived.
> 2. Similarly, those quantities that cannot be analytically composed from given quantities analytical one with the other are not themselves analytical with respect to the quantities in question.[39]

The first postulate tells us that if a and b are mutually commensurable magnitudes, then $c = a + b$, or $d = a^3 b$, $e = \sqrt[3]{ab}$ and all of the quantities likewise composed from a and b via finitely many analytical operations are all analytical one with respect to the other. This also implies that the relation of "being analytical" is a transitive relation, besides being symmetric and reflexive. In fact, let us suppose that a is analytical with respect to b, and b is analytical with respect to c. By definition, b is composed from a by a finite combination of analytical or arithmetic operations, and c is likewise obtained from b by a finite combination of analytical compositions. Thus, c can be obtained by a via a finite combination of analytical compositions as well. Hence, c is analytical with respect to a in virtue of the first postulate. Therefore, the relation of "being analytical" is an equivalence relation.

The second postulate, which introduces non-analytical quantities in Gregory's theory, enables us to assert that a quantity is analytical with respect to given analytical quantities if and only if it can be analytically composed from them. The "'only if" part should be stressed: we could say that, by this choice of postulates, Gregory aimed to characterize analytical quantities by the fact of their being closed or stable with respect to the five arithmetic operations. This is a fundamental insight laying the grounds for Gregory's impossibility argument, because it introduces the class of Cartesian, or algebraic, quantities as the proper framework with respect to which the impossibility of solving the quadrature of the central conic section should be understood.

If we then imagine choosing a reference quantity C, a partition can be introduced within the set of quantities that defines two classes $E_c = \{D/analytical(D, C)\}$, where D is another, arbitrary quantity, and $F_c = \{A/non - analytical(A, C)\}$, where A is also an arbitrary quantity. The language employed here is modern, but the fundamental idea is already present in Euclid's characterisation of commensurable and incommensurable quantities, which Gregory must have known.[40] In fact, an analogy can be drawn with Propositions 12 and 13 of Euclid's Book X, in which the relation 'being commensurable'

---

[39] *VCHQ*, p. 10: "1. Petimus quantitates, a quantitatibus datis inter se analyticis analytice compositas, esse inter se & cum quantitatibus datis analyticas. 2. Item quantitates, quae quantitatibus dates inter se analyticas non possunt analytice componi, non esse cum quantitatibus datis analyticas."

[40] See Scriba (1957, p. 14).

is proven to be an equivalent relation and, on the strength of this result, the domain of quantities can be partitioned into two disjoint sets (see Vitrac's commentary in Euclid (1990, Vol. 3, p. 135)). However, unlike incommensurable quantities, Gregory was not yet able to exhibit any non-analytical quantity. The reason is obvious, since the existence of non-analytical quantities would be proven by him only later in the book, with the impossibility of the analytical quadrature of the central conic sections. The fact that, at the beginning of the *VCHQ*, Gregory could not give any example of a non-analytical quantity or a non-analytical composition explains the necessity of postulating their existence, in an hypothetical way, via the above demands.

A good candidate for a non-analytical composition is represented by the operation of forming "convergent series" (*series convergentes*),[41] defined by Gregory as follows:

Definition 2.1 (*VCHQ*, df. 9, p. 10) Given two successions of quantities $\{a_n\}$ and $\{b_n\}$, Gregory called "*convergent series*" a double sequence $\{a_n, b_b\}$, if the following conditions are obtained (S and S' are two finite compositions):

There is a composition S such that

$$\forall n, a_{n+1} = S(a_n, b_n).$$

There is a composition S' such that

$$\forall n, b_{n+1} = S'(a_n, b_n)$$

and:

$$\forall n \mid b_{n+1} - a_{n+1} \mid < \mid b_n - a_n \mid .$$

Gregory's definition is abstract, but, in practice. it captures the behaviour of one special infinite convergent series related to the area of a conic sector, namely, the series $\{I_n, C_n\}$ of polygonal approximations, as I will explain in the next section.

Gregory's definition expresses in formal terms, without involving any idea of a limit-quantity, the intuitive fact that the terms (a_n, b_n) of a convergent series approach each other more and more as n grows. Gregory's understanding and use of the term "convergent" differ from our own. For instance, the above definition does not explicitly rule out the case of two sequences $\{a_n\}$ and $\{b_n\}$ of quantities or numbers that grow closer and closer to

[41] In the rest of the chapter, I shall use, according to Gregory's terminology, the expression "convergent series." As the subsequent discussion will clarify, the term "convergent" is employed here in a slightly different sense than the modern one. Therefore, in the following sections, any reference to the modern meaning of convergence will be made explicit, when necessary, in order to avoid ambiguities.

each other, whereas the sequence of the differences does not tend to 0. This being said, as we shall see in the next section, all of the convergent series discussed in the *VCHQ* are such that the sequence of the differences tends to 0, and their limit always exists on the grounds of the assumed continuity of geometrical space.

2.3.3 The Convergent Series of Polygonal Approximations

In the first six Propositions of *VCHQ*, Gregory set out to prove that the double sequence of the in- and circumscribed polygons to an arbitrary sector of a central conic section forms a convergent series in the precise sense stated in Definition 9, when the polygons are constructed in the protocol succinctly described in the previous section.

This is first done by proving, in Propositions I-V of the *VCHQ*, the existence of two analytical compositions that relate the areas of any pair of in- and circumscribed polygons (in symbols: $\{I_n, C_n\}$) to the areas of the next couple of polygons (in symbols: $\{I_{n+1}, C_{n+1}\}$). Then, in Proposition VI, Gregory proves that the series of polygons is a convergent series according to the definition given in the *VCHQ*.

In this section, I shall discuss in detail both steps of Gregory's procedure. Gregory's argument begins with the three following theorems (corresponding to Propositions I, II and III of the *VCHQ*):

Theorem 2.1 (*VCHQ*, Proposition 1) *With reference to Fig. 2.4, let ABP be an inscribed triangle determined by the centre A of the conic and by two points B and P lying on the conic, let $ABFP$ be a quadrilateral delimited by points A, B, P and by a fourth point F, the intersection of the tangents BF and PF to the conic, and let $ABIP$ be an inscribed quadrilateral delimited by points A, B, P and I, the intersection between FA and the conic section. Then, the following proportion holds: $ABFP : ABIP = ABIP : ABP$.*

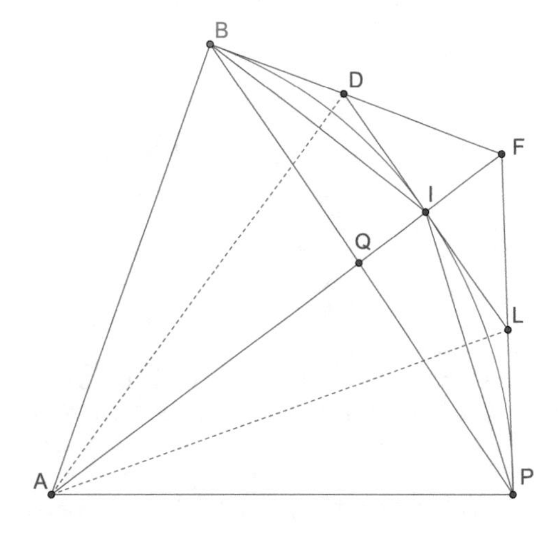

Fig. 2.4 A sector of the ellipse, approximated by the inscribed triangle ABP, the circumscribed quadrilateral $ABFP$ (first pair), and by an inscribed quadrilateral $ABIP$ and a circumscribed pentagon $ABDLP$ (second pair)

Theorem 2.2 (*VCHQ*, Proposition 2 and 3) *Let us consider polygons ABP and $ABIP$, as in the previous theorem, and let $ABDLP$ be the pentagon formed by the already marked points A, B, P and by D and L, intersections between the tangent DL to the conic sector at I and the tangents to the conic at P and B. Then, the following proportions hold between polygons ABP, $ABIP$ and $ABDLP$:*

$$(ABP + ABIP) : ABIP = ABFP : ABDLP;$$

$$(ABP + ABIP) : ABIP = 2ABIP : ABDLP.$$

Gregory proved that which I refer to here as Theorems 2.1 and 2.2 relying on the theory of proportions and on the geometry of conic sections. In order to familiarise ourselves with Gregory's classical method, it is thus worth reconstructing both proofs of the above theorems.

Let us then begin with the proof of Theorem 2.1. Gregory used the following projective property: segment AI is the mean proportional between AQ and AF.[42]

Since triangles ABF, ABI and ABQ lie on the same line and have the same height, in virtue of *Elements*, VI, 1, they also stand in the same proportion as their basis, so that the following proportions hold:

$$ABQ : ABI = AQ : AI = AI : AF = ABI : ABF.$$

Since we have, by construction, that ABQ is half of ABP, ABI half of $ABIP$ and ABF half of $ABFP$, we shall also have that ABP, $ABIP$ and $ABFP$ are in continuous proportion, namely: $ABFP : ABIP = ABIP : ABP$, which proves Theorem 2.1.

In order to prove Theorem 2.2, Gregory began with the following proportions, which can be easily proven using *Elements*, VI, 1:

$$ABFP : ABIP = AF : AI$$

and

$$ALF : ALI = AF : AI.$$

[42] The proof can be easily recovered relying on *Conica*, I, 37, because the line passing through points A, Q, I and F is a diameter of the section (since it passes through the center), and the segment PB is an ordinate in the terminology of Apollonius' theory of conic sections. Indeed, according to Apollonius' *Conica*, I. 37: "In a hyperbola, an ellipse or a circle, if QV be an ordinate to the diameter PP', and the tangent at Q meets PP' in T, then: $CV.CT = CP^2$" (Heath 1896, p. 28).

Therefore, we can conclude that:

$$ABFP : ABIP = ALF : ALI.$$

The above proportion yields, *componendo*, and, doubling the second and fourth term, the following:

$$(ABFP + ABIP) : 2ABIP = AFP : 2ALI. \tag{2.3}$$

On the other hand, an inspection of Fig. 2.4 shows that the following equalities hold by construction:

$$2ALI = AILP = \frac{ABDLP}{2}.$$

$$AFP = \frac{ABFP}{2}.$$

On this basis, we derive, from (2.3),

$$(ABFP + ABIP) : 2ABIP = ABFP : ABDLP. \tag{2.4}$$

But this is the first proportion stated in Theorem 2.2.

In order to prove the second proportion, let us proceed as follows. From the above (2.4) and from Theorem 2.1, we can infer that[43]

$$(ABP + ABIP) : ABIP = 2ABIP : ABDLP.$$

Again, from Theorem 2.1, we obtain, *componendo*,

$$(ABIP + ABFP) : ABFP = (ABP + ABIP) : ABIP.$$

Meanwhile, from (2.4), we have, *permutando*,

$$(ABIP + ABFP) : ABFP = 2ABIP : ABDLP.$$

Comparing the last two proportions obtained, we infer that

$$(ABP + ABIP) : ABIP = 2ABIP : ABDLP,$$

which completes the proof of Theorem 2.2.

[43]Let us start from the proportion: $ABFP : ABIP = ABIP : ABP$. *Invertendo*, we obtain: $ABIP : ABFP = ABP : ABIP$, and *componendo*: $(ABIP + ABFP) : ABFP = (ABP + ABIP) : ABIP$. If we now consider (2.4), we have, *permutando*: $(ABFP + ABIP) : ABFP = 2ABIP : ABDLP$. We can now compare the proportion: $(ABIP + ABFP) : ABFP = (ABP + ABIP) : ABIP$ and $(ABFP + ABIP) : ABFP = 2ABIP : ABDLP$, since their left-hand sides are equal.

In Propositions IV and V, Gregory generalized Theorems 2.1 and 2.2 to the subsequent terms of the series, and concluded, in the *Scholium* of Proposition V, that they could be generalized to any pair of successive terms, on the basis of the geometrical recursive construction that generates the double sequence of in- and circumscribed polygons:

> The two previous propositions can be proven in the same way for any two in- and circumscribed polygons besides the in- and circumscribed polygons $ABIP$ and $ABDLP$; in fact, the polygon included by the tangents contains a great many many equal trapezia, as many equal triangles as are contained in the polygon included under the chords.[44]

Having proven purely geometrically that the same relations of proportionality hold between each pair of polygons and the subsequent one, Gregory could eventually define a (convergent) series, complying with Definition 9 of the *VCHQ*.

For this purpose, he relied on the symbolism of Cartesian algebra in order to denote the areas of in- and circumscribed polygons, and it was in this way that he interpreted the proportions appearing in Theorems 2.1 and 2.2.[45]

Thus, as we read in the *Scholium* of Proposition V, Gregory posited $ABP = a$, $ABFP = b$, and deduced: $ABIP = \sqrt{ab}$ (let us recall that $ABIP$ is the mean proportional between ABP and $ABFP$). In a similar way, he derived, relying on Propositions II and III of the *VCHQ* (cf. Theorem 2.2): $ABDLP = \frac{2ab}{a+\sqrt{ab}}$, and by simple substitution, he concluded that the same analytical relations held between successive pairs of polygons.

In order to emphasise the structure of Gregory's series, I will employ the symbols: 'I_0', 'C_0' $\ldots I_n, C_n \ldots$, starting from $ABP = I_0$, $ABFP = C_0$, $ABIP = I_1$ and $ABDLP = C_1$. Thus, the first proportion, namely, $ABFP : ABIP = ABIP : ABP$, will yield

$$I_1 = \sqrt{C_0 I_0}.$$

The second one, namely, $(ABP + ABIP) : ABIP = 2ABIP : ABDLP$, will yield

$$C_1 = \frac{2C_0 I_0}{I_0 + \sqrt{C_0 I_0}}.$$

[44]*Scholium* to Prop. V: "Duae praecedentes propositiones eodem modo demonstrari possunt de duobus quibuscumque polygonis complicatis loco polygonorum complicatorum $ABIP$, $ABDLP$; polygonum etiam a tangentibus comprehensum tot continet aequalia trapezia, quot continet polygonum a subtendentibus comprehensum."

[45]Cf., for instance, the *Scholium* to Proposition V, p. 15.

Generalizing Theorems 2.1 and 2.2, we arrive at the conclusion that each pair $\{I_n, C_n\}$ is related to the next one $\{I_{n+1}, C_{n+1}\}$ by a pair of analytic compositions. This result can be summed up in the following theorem:

Theorem 2.3 *For two arbitrary pairs* (I_n, C_n) *and* (I_{n+1}, C_{n+1}) *of in- and circumscribed polygons to a conic sector, the following relations hold:*

$$I_{n+1} = \sqrt{C_n I_n},$$

$$C_{n+1} = \frac{2C_n I_n}{I_n + \sqrt{C_n I_n}}.$$

We note that the series $\{I_n, C_n\}$ is formed by quantities analytical one to the other if and only if the initial terms are analytical one to the other.

Considering, for the sake of simplicity, just the case of a circular sector with radius r bounded by an arc θ (Fig. 2.5), we can express the areas of the in- and circumscribed polygons using trigonometry.

Thus, we shall have that

$$ABP = r^2 sin(\frac{\theta}{2}) cos(\frac{\theta}{2}),$$

$$ABFP = r^2 tan(\frac{\theta}{2}).$$

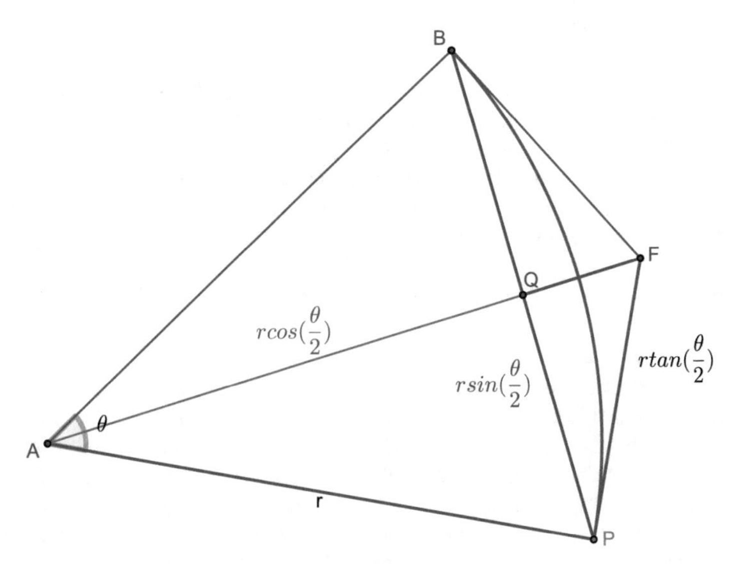

Fig. 2.5 Computing the area of polygons, using trigonometry. $AQ = rcos(\frac{\theta}{2})$, $QP = rsin(\frac{\theta}{2})$, $FP = rtan(\frac{\theta}{2})$

The above representation makes it clear that, in the case of the circle, the areas of these polygons are analytically composed from the radius of the circle and the chord corresponding to the chosen sector, which measures $2rsin(\frac{\theta}{2})$. Moreover, if the radius and the chord are analytical with respect to one another, then the areas of ABP and $ABFP$ will be analytical with respect to one another (in virtue of Postulate 1).This result is important, because, as I shall mention later on (Sect. 2.4.2), Gregory infers the impossibility of squaring the whole circle ("definite" quadrature) from this result.

As a final comment on Gregory's mathematical practice, it is worth stressing the importance that the use of symbolic algebra acquired for him in order to extrapolate an analytical convergent series from the double succession of polygons. In fact, solely relying on manipulation of proportions between polygonal areas, without the aid of symbolic manipulations, it would have been very difficult to single out the recursive relations between successive polygons. Thus, it would have been extremely difficult to identify them as an example of a convergent series.

In order to compute his series, Gregory relied on the use of algebraic symbols to denote geometrical quantities, in this case, the areas of the in- and circumscribed polygons, as well as the area of the sector itself. Gregory's practice reminds us of Descartes' use of algebra in geometry, according to the method introduced in the first book of the *Géométrie*. Even if Gregory's use of the symbolic language of algebra did not seem to closely follow Descartes' mathematical practice, as letters are merely employed, in the *Géométrie*, to denote known and unknown segments and never for areas, such a use was admitted in practice among Cartesian geometers. For instance, the Cartesian mathematician F. De Beaune made the point explicit in his short tract *Notae Breves*, appended to the first Latin edition of the *Géométrie*:

> But in order to establish the precepts of this science and pursue its knowledge, it is very good to consider, in general, ratios in terms of segments. Indeed they are very simple, and they vindicate for themselves the expression of all the ratios, which can be considered among any other quantities. Numbers are unable to do this, since they cannot express relations that are found among incommensurable quantities. Add that we can use them for all the other things, which have a mutual ratio or a proportion ... hence we can express by means of two lines the ratio between two surfaces, or between two different velocities, and between other things of the same kind, which have some relation between them.[46]

[46]The passage comes from De Beaune's *In Geometriam Renati Descartes Notae Breves*, published as an appendix to the first Latin edition of the *Géométrie*. See Descartes (1659, p. 108): "Optimum verum est, ad stabilienda huius Scientiae praecepta, et ad cognitionem ejus assequendam, ut generaliter rationes hasce in lineis consideremus: cum simplicissimae sint, et hoc sibi vindicent, quod rationes omnes, quae inter quascumque alias res considerari possunt, exprimant. Id quod numeri non efficiunt, qui relationes, quae inter incommensurabiles quantitates reperiuntur, exprimere nequeunt. Accedit, quod iis ad omnes alias res, rationem vel proportionem quandam inter se habentes, uti possumus ... possumus tamen rationem quae inter duas superficies, aut inter duas differentes velocitates, et id genus alia, quae inter se relationem aliquam habere statuimus, reperitur, exprimere per duas lineas."

The science to which De Beaune is here referring is Descartes' algebra, which the latter also referred to as a "calculus" of segments. In De Beaune's opinion, Descartes' algebra was a symbolic system that could denote various classes of quantities, provided dimensional homogeneity was respected. Significantly, two examples are given in the passage above: velocities, which come from physics, and surfaces, which come from geometry. The latter example is particularly relevant for our present inquiry, since it provides a historical justification for Gregory's practice of denoting polygonal magnitudes by letters.

Thus, Gregory's symbolism in such expressions as "$ABP = a$" can be interpreted in the following sense: "the area of the polygon ABP is measured by the segment a in the domain of segments constructed from an arbitrarily chosen unity segment." Analogously, "$ABIP = \sqrt{ab}$" expresses the fact that the area of the figure $ABIP$ is measured by the segment $\sqrt{ab}$ within the same domain of segments. By virtue of the geometrical interpretation of multiplication in force within Descartes' geometry, moreover, even more complex expressions can still be interpreted as segments in the plane.

Thanks to the use of algebra, the problem of squaring a sector of the circle (the ellipse or the hyperbola) would become, in Gregory's understanding, the problem of constructing a segment that measures, in the domain of segments, the area of the sector once having fixed a unit of measurement. Moreover, the series of segments measuring the areas of in- and circumscribed polygons is convergent, namely: (i) each term (a segment, as per our interpretation) can be analytically constructed from its predecessors, (ii) the difference between each couple (I_n, C_n) of corresponding terms grows smaller as the index n increases.

While I have discussed in this section how Gregory proved the former property (i), in the next one, I shall turn to Gregory's proof of (ii).

2.3.4 Proving the Convergence of the Series

In order to prove the convergence of the series, in the sense of the definition given in the preface of the *VCHQ*, Gregory showed that the differences between the successive terms of the series $\{I_n\}$ and $\{C_n\}$, as n increases, become smaller, according to the following relation: $\forall n, (C_{n+1} - I_{n+1}) < \frac{1}{2}(C_n - I_n)$ in the case of the circle or the ellipse, and $\forall n (I_{n+1} - C_{n+1}) < \frac{1}{2}(I_n - C_n)$ in the case of the hyperbola.

Considering, for the sake of simplicity, only the case of the series approaching a sector of the circle or of an ellipse (in the case of the hyperbola, the following inequalities must be inverted), Gregory proceeded by giving a direct proof of the following inequality for the first pair of in- and circumscribed polygons:

$$C_1 - I_1 < \frac{1}{2}(C_0 - I_0). \tag{2.5}$$

He then generalized this inequality to successive terms, on the strength of the recursive polygonal construction.[47]

Let us therefore reconstruct Gregory's proof of the inequality labelled (2.5). The proof begins from the proportions proven in Theorems 2.1 and 2.2 above.

In symbols, these proportions become

$$I_0 : I_1 = I_1 : C_0 \tag{2.6}$$

and

$$(I_0 + I_1) : I_1 = 2I_1 : C_1. \tag{2.7}$$

From (2.6), one derives, through elementary manipulations, the following[48]:

$$(C_0 - I_0) : (C_0 - I_1) = (I_0 + I_1) : I_1. \tag{2.8}$$

Moreover, (2.7) yields, *permutando*: $(I_0 + I_1) : 2I_1 = I_1 : C_1$, and then $(I_1 - I_0) : 2I_1 = (C_1 - I_1) : C_1$ (*invertendo* and *separando*). From which it follows, *invertendo* again:

$$(I_1 - I_0) : (C_1 - I_1) = 2I_1 : C_1. \tag{2.9}$$

We thus have, from (2.7) and (2.8), the following:

$$(C_0 - I_0) : (C_0 - I_1) = 2I_1 : C_1.$$

From this proportion, and from (2.9), we obtain

$$(C_0 - I_0) : (C_0 - I_1) = (I_1 - I_0) : (C_1 - I_1). \tag{2.10}$$

Since $I_1 > I_0$ holds by construction, Gregory inferred the new inequality:

$$(C_0 - I_0) > (C_0 - I_1).$$

Moreover, by virtue of proportion (2.10), Gregory also concluded the following:

$$(I_1 - I_0) > (C_1 - I_1).$$

[47] *VCHQ*, Prop. VI and in the succeeding *scholium*, pp. 16–18.

[48] From $I_0 : I_1 = I_1 : C_0$, one obtains, *invertendo*, *separando* and *permutando*: $(I_1 - I_0) : (C_0 - I_1) = I_0 : I_1$. The last proportion yields, *componendo*: $(C_0 - I_0) : (C_0 - I_1) = (I_0 + I_1) : I_1$.

Permutando, proportion (2.10) yields this new proportion:

$$(C_0 - I_0) : (I_1 - I_0) = (C_0 - I_1) : (C_1 - I_1).$$

Since $I_0 < I_1 < C_0$ holds by construction, the following two inequalities will hold too (their proof is obvious, given the definition of proportional magnitudes):

$$(C_0 - I_0) > (I_1 - I_0), (C_0 - I_1) > (C_1 - I_1).$$

From $(C_0 - I_1) > (C_1 - I_1)$ and from $(I_1 - I_0) > (C_1 - I_1)$ (the last inequality is proven above), Gregory could finally infer that

$$(C_0 - I_0) > 2(C_1 - I_1).$$

Or, in Gregory's words: "the difference between triangle ABP and trapezoid $ABFP$ is greater by double the difference between the trapezoid $ABIP$ and the polygon $ABDLP$."[49]

In the *Scholium* that immediately follows Proposition 6, Gregory generalized the above inequality to any successive term of the series on the basis of the geometrical recursive procedure, so that he obtained the following, general result (abbreviated using a modern formalism)[50]:

$$\forall n, 2(C_{n+1} - I_{n+1}) < (C_n - I_n). \tag{2.11}$$

Therefore, the series $\{I_n, C_n\}$ converges according to the definition given in df. 9 of the *VCHQ*. Gregory also remarked that the series of the differences can be made smaller than any assigned quantity:

> For all the polygons described to the infinite ... the difference of the former pair of inscribed and circumscribed ones will always be greater than the double of the difference of those inscribed and circumscribed polygons immediately following in the series, in such a way that in order to make the next difference one has to subtract from the previous difference more than its half, and thus continuing the construction of the subduplicate series of polygons, a pair of in- and circumscribed polygons can be found, whose difference is less than any exhibited quantity.[51]

---

[49] *VCHQ*, p. 16.

[50] *VCHQ*, schol. 6, p. 17.

[51] *VCHQ*, p. 17: "in ... nostra polygonorum complicatorum descriptione in infinitum; differentia enim priorum nempe inscripti et circumscripti maior semper erit duplo differentiae immediate sequentium nimirum inscripti quoque et circumscripti, et proinde aufertur maius quam dimidium a priorum differentia ut fiat differentia immediate sequentium; et igitur continuando subdupla

Since Gregory's method does not admit of infinitesimal magnitudes, from the inequality stated in Theorem 2.11, it follows, using a slightly anachronistic terminology, that:

$$\lim_{n\to\infty} |I_n - C_n| = 0.$$

I point out that, in order to carry out his proof, Gregory considered the differences between two polygonal areas as objects themselves. This would lead to a substantial simplification of the classical indirect methods based on double reduction, as we shall also see below in the proof that the series of polygons converges to the area of the sector.

As the series of the differences converges to zero, Gregory concludes that there exists a difference that can be denoted as $|I_\infty - C_\infty|$, which terminates it. Gregory called the pair (I_∞, C_∞) "*terminatio*," an expression that we can translate as the "termination" or "limit" of the series of in- and circumscribed polygons.[52]

In the *VCHQ*, Gregory took for granted that the termination $I_\infty (= C_\infty)$ is identical to the area of the sector, and therefore reduced the problem of squaring the latter to the problem of computing the termination itself. In his second published work, the *Geometriae pars universalis*, Gregory supplemented a proof that the limit of the series of in- and circumscribed polygons is the sought-for area of the sector:

> There are some who claim that I have not well proven (in the *scholium* of Proposition 5) that the sector $A\overset{\frown}{BI}P$ [in our account, $A\overset{\frown}{BP}$] is the same as the limit of the convergent series, whose first terms are the triangle ABP and the trapezium $ABFP$, and whose second terms are the polygons $ABIP$ and $ABDLP$. And therefore I submit here a full demonstration.[53]

Gregory's full demonstration proceeds as follows. Let us call θ the area of the given sector. For the sake of contradiction, let us suppose that the limit of the convergent series differs from the said area, namely: $|I_\infty - \theta| \neq 0$. Hence, this difference will be equal to a quantity Z.

Let us then construct a pair of polygons (I_n, C_n), such that $|I_n - C_n| < Z$.[54]

polygonorum descriptionem, inveniri possunt duo polygona complicata, cujus differentia sit minor qualibet exhibita quantitate."

[52]Cf. *VCHQ*, p. 19: "...imaginando hanc seriem in infinitum continuari, possumus imaginari ultimos terminos convergentes esse aequales, quos terminos aequales appellamus seriei terminationem".

[53]*GPU*, pref., p. 6: "Dicunt alii non bene esse demonstratum (in scholium prop. 5) sectorem $ABIP$ eundem esse cum terminationem seriei convergentis, cuius primi termini sunt triangulum ABP et trapezium $ABFP$, et secundi, rectilinea $ABIP$, $ABDLP$." I have not been able to detect the source of the critiques mentioned in the passage.

[54]This is possible by virtue of the convergence of the series, as stated in *VCHQ*: "Differentia enim polygonorum complicatorum in seriei continuatione sempre diminuitur, ita ut omni exhibita quantitate fieri possit minor" (*VCHQ*, *Scholium* V, p. 15).

Hence, $|I_n - C_n| < |I_\infty - \theta|$. However, by construction of the series, we have either that: $I_c < I_\infty < \theta < C_n$ or $I_c < \theta < I_\infty < C_n$. In either case, a contradiction arises.

The problem of finding the area of a sector of a central conic section was thus reduced to that of computing or constructing the limit of its convergent polygonal series. This was certainly not an easy task, as we read in the *VCHQ*:

> If, then, the said sequence of polygons could be terminated, that is to say, if it were found that the last inscribed polygon (if I may say so) were equal to the last circumscribed polygon, we would have without fail the quadrature of the circle and the hyperbola. But since it is difficult in geometry, and perhaps cannot even be imagined to be possible to terminate this series, we have to premise some propositions from which the terminations of such series can be found, and finally (if it can be done) a general method for the discovery of the terminations of all convergent series.[55]

2.3.5 Computing the Limit of a Series

A parallel can be draw between the two styles of analysis performed by Descartes and Gregory, for they both focus on reducing a geometrical problem to a problem of algebra. Cartesian analysis proceeds by reducing a geometrical problem to the problem of solving a finite algebraic equation, whereas Gregory's analysis reduces the geometrical problem of squaring the sector of a central conic to the problem of finding the limit of a convergent series. Just like Descartes devised a method to solve equations geometrically by constructing the unknown through the intersection of curves, the corresponding move undertaken by Gregory in the *VCHQ* was to devise a general procedure in order to compute the limit of a convergent sequence. The motive for this procedure was the hope that the most difficult cases, like the one related to the quadrature of the central conic sections, would fall under such a method.[56]

As is evinced by Propositions VII-IX of the *VCHQ*, Gregory first proceeded by discussing techniques that worked for specific series and then generalized these examples

[55] *VCHQ*, *Scholium* to proposition V, p. 15: "Si igitur praedicta polygonorum series terminari posset, hoc est, si inveniretur ultimum illud polygonum inscriptum (ita loqui licet) aequale ultimo illo polygono circumscripto, daretur infallibiliter circuli et hyperbolae quadratura: sed quoniam difficile est in geometria, omnino fortasse inauditu tales series terminare, praemittendae sunt quaedam propositiones e quibus inveniri possint huiusmodi aliquot serierum terminationes, et tandem (si fieri possit) generalis methodus inveniendi omnium serierum convergentium terminationes."

[56] *VCHQ*, pp. 19–24. See also Dehn and Hellinger (1939, pp. 472–473) and Scriba (1957, pp. 16–17).

into a set of rules for the computation of the limit of a generic convergent series, in Proposition X:

> (Proposition X) From a given quantity, composed in the same way from any two terms of any convergent series and from the two immediately successive terms of the same series, to find the termination of the proposed series.[57]

Gregory's method can be explained as follows. Let us start from a pair of convergent sequences of this form (with M and M' analytical compositions):

$$a_{n+1} = M(a_n, b_n),$$
$$b_{n+1} = M'(a_n, b_n).$$

On Gregory's understanding, the problem of computing the limit can be reduced to the problem of finding a certain analytical composition f such that: $f(a_n, b_n) = f(a_{n+1}, b_{n+1})$, for every n. Calling the termination z, if such a composition f exists, we obtain the following:

$$f(a_0, b_0) = f(a_1, b_1) = \ldots = f(a_n, b_n) = \ldots = f(a_{n+1}, b_{n+1}) = \ldots = K,$$

where K is a constant, by definition analytical with the terms of the series. The termination of the series can then be computed by solving the following equation, in the unknown z:

$$f(z, z) = K. \quad (2.12)$$

Since the composition f is, by hypothesis, invariant for any two arbitrary pairs (a_n, b_n), (a_{n+1}, b_{n+1}), it will satisfy the (polynomial) equation $f(a_0, b_0) = f(a_1, b_1) = K$ too, hence K can be calculated from the first and second convergent terms, once f has been determined:

> Conclusively, in order to find the termination of any convergent sequence, one only needs to find a quantity composed in the same way from the first convergent terms as from the second convergent terms.[58]

[57] "Ex data quantitate, eodem modo composita a duobus terminis convergentibus cujuscumque seriei convergentis, quo componitur ex terminis convergentibus ejusdem seriei immediate sequentibus, seriei propositae terminationem invenire." (*VCHQ*, p. 23).

[58] *VCHQ*, p. 24: "Et proinde ad inveniendam cujuscumque seriei convergentis terminationem; opus est solummodo invenire quantitatem eodem modo compositam ex terminis convergentibus primis quo componitur eadem quantitas ex terminis convergentibus secundis."

We can gain a better grasp of Gregory's procedure if we examine an example presented in Gregory's response to a criticism raised by Huygens from July 1668.[59]

In the letter, Huygens had pointed out that Gregory committed a mistake by considering a non-convergent series as if it were convergent. Gregory acknowledged the correctness of Huygens' criticism and amended his previous faulty example by proposing another pair of sequences, for which he (correctly) assumed their convergence. The pair chosen by Gregory, which we may denote by: $\{a_n, b_n\}$ with $0 < a_n < b_n$, is defined recursively via the following conditions:

$$a_0 = a, \tag{2.13}$$

$$b_0 = b \tag{2.14}$$

and

$$a_{n+1} = \frac{2a_n b_n}{a_n + b_n}, \tag{2.15}$$

$$b_{n+1} = \frac{a_n + b_n}{2}. \tag{2.16}$$

The convergence of this two-term recursion can be easily proven. The conditions of convergence established by Gregory are satisfied, since the following inequalities hold:

$$a_n < \frac{2a_n b_n}{a_n + b_n} < b_n,$$

$$a_n < \frac{a_n + b_n}{2} < b_n.$$

From them, it can be directly concluded that

$$\frac{a_n + b_n}{2} - \frac{2a_n b_n}{a_n + b_n} < b_n - a_n.$$

Thus, for any n, we must conclude that

$$b_{n+1} - a_{n+1} < b_n - a_n.$$

Gregory assumed that this sequence had a single termination (the proof that the series also converges in our modern sense can be easily supplied), analytical with respect to its

[59] See Gregory (1668a) and Huygens (1888–1950, Vol 6, p. 229).

terms, and calculated it. By virtue of the recursive nature of the sequences, an arbitrary pair (a_k, b_k) of terms, $a_k \in \{a_n\}$ and $b_k \in \{b_n\}$, can be written as

$$a_k = \frac{2a_{k-1}b_{k-1}}{a_{k-1} + b_{k-1}},$$
$$b_k = \frac{a_{k-1} + b_{k-1}}{2}.$$

From which it results that the product of a_k and b_k equals the product of a_{k-1}, b_{k-1}:

$$a_k b_k = (\frac{2a_{k-1}b_{k-1}}{a_{k-1} + b_{k-1}})(\frac{a_{k-1} + b_{k-1}}{2}) = a_{k-1}b_{k-1}.$$

As a consequence of the recursive nature of the successions $\{a_n\}$, $\{b_n\}$, we shall find that

$$a_k b_k = a_{k-1}b_{k-1} = \ldots = a_1 b_1 = ab.$$

Therefore, the product of the terms a_k and b_k is invariant for any choice of the index k. From this, Gregory concluded that, for any pair (a_k, b_k), the product of its terms equals the product of the known initial terms a_0 and b_0:

> The first terms multiplied one by the other yield *ab*, then the next ones multiplied one by the other give the same *ab*, from them the limit of the proposed *series* must be discovered. It is plain that the quantity *ab* is composed in the same way from the convergent terms *a* and *b* as from the convergent terms immediately following, $\frac{2ab}{a+b}$, $\frac{a+b}{2}$; and since quantities *a* and *b* can stand for any term of the convergent sequence, indefinitely, it is evident that any two convergent terms of the proposed series multiplied one by the other yield the same product, obtained from the immediately following terms multiplied one by the other; and since two convergent terms are always immediately followed by two other convergent terms, it is plain that two arbitrary convergent terms multiplied by one another always yield the same product, namely *ab*.[60]

[60]Huygens (1888–1950, Vol VI, pp. 241–242): "Termini priores inter se multiplicati efficiunt *ab*, item sequentes inter se multiplicati efficiunt eandem *ab*, ex his invenienda sit propositae seriei terminatio. Manifestum est, quantitatem *ab* eodem modo fieri a terminis convergentibus *a*, *b*, quo a terminis convergentibus immediate sequentibus $\frac{2ab}{a+b}$, $\frac{a+b}{2}$: & quoniam quantitates *a*, *b*, indefinite ponuntur pro quibuslibet totius seriei terminis convergentibus, evidens est, duos quoscunque terminos convergentes propositae seriei inter se multiplicatos idem efficere productum, quod faciunt termini immediate sequentes etiam inter se multiplicati; cumque duo termini convergentes duos terminos convergentes semper immediate sequantur, manifestum est, duos quoscunque terminos convergentes inter se multiplicatos idem semper efficere productum, nempe *ab*."

Following the reasoning expounded in the letter, the recursive character of the sequences under examination allows us to establish the following equality:

$$z.z = \ldots = a_k b_k = a_{k-1} b_{k-1} = \ldots = a_1 b_1 = ab.$$

From this, Gregory was able to derive a quadratic equation in z:

$$z.z = z^2 = ab.$$

The problem of expressing z in terms of a and b (both assumed to be positive) can be solved arithmetically, or it can be solved geometrically if z is taken to be a line segment. In the end, the termination of this sequence is: $z = \sqrt{ab}$, namely, the geometrical mean of the terms a, b.[61]

Let us consider once again the logical structure of Gregory's method for computing the termination of the series. In the examples considered, we begin by supposing that there exists an invariant composition for all of the terms of the series, and conclude that the termination can be computed accordingly. Spurred by Huygens' and Wallis's objections, Gregory also claimed the converse, namely, that if the termination is analytical with the series, then it could be computed using the method illustrated in this section. This point stood at the core of the dispute between Gregory and Huygens, and I will come back to it later.

2.4 The Taming of the Impossible

2.4.1 An Argument of Impossibility

Gregory's impossibility result relied on a tacit assumption, which we can state in these terms: if a series converges to an analytical limit, this limit can be computed using the method expounded in Propositions VII-X of the *VCHQ*. Huygens would criticize this assumption in the report that appeared in the *Journal des Sçavans* at the beginning of July 1668, as will be discussed below.

On the basis of this assumption, Gregory applied the above method to computing the limit of the double succession $\{I_n\}$ and $\{C_n\}$ of in- and circumscribed polygons to an arbitrary conic sector $\widehat{APB}$.

As recalled in the preface of the *VCHQ*, this attempt was eventually unsuccessful, due to being hindered by "unsurpassable difficulties." These difficulties, however, would turn

[61] See Dehn and Hellinger (1939, p. 473) and Borwein (1998, p. 4).

out to be rich in consequences, since they led to the central result of the *VCHQ*, which can be phrased in these terms[62]:

Theorem 2.4 *The area of a sector of the circle, ellipse, or hyperbola* $\widehat{APB}$ *(indefinite quadrature) cannot be composed analytically from the areas of the inscribed triangle* ABP *and the circumscribed trapezium* $ABFP$*, analytical with respect to one another.*

Gregory proved the above theorem by a *reductio* argument. He started by assuming that the area of the sector $\widehat{APB}$ could be analytically composed from the areas of the in- and circumscribed polygons ABP and $ABFP$. This being granted, the area of $\widehat{APB}$ would be analytical with respect to the areas of the polygons ABP and $ABFP$. Therefore, on the basis of the procedure described in the previous section, the limit of the convergent series could be computed by solving the equation: $S(I_0, C_0) = S(I_\infty, C_\infty) = K$, where the polygons (I_∞, C_∞) are the limit of the series, and S is an analytical composition of I_i and C_i, such that $S(I_n, C_n) = S(I_{n+1}, C_{n+1}) = S(I_\infty, C_\infty)$. As we know that $(I_\infty = C_\infty)$, we can write the previous chain of equalities as: $S(I_n, C_n) = S(I_{n+1}, C_{n+1}) = S(I_\infty, I_\infty)$

On the other hand, if the limit could be expressed analytically in terms of I_0 and C_0, it could be expressed by the same analytical composition S in terms of I_1 and C_1, so that the following identity would hold:

$$S(I_0, C_0) = S(I_1, C_1) = S(\sqrt{I_0 C_0}, \frac{2 I_0 C_0}{I_0 + \sqrt{I_0 C_0}}). \tag{2.17}$$

Thus, it would be sufficient to argue that no analytical composition S satisfying the above identity exists in order to conclude that the limit ϖ is not analytical with the terms of the sequences and, by contradiction, that the sector could not be squared analytically.

For the sake of contradiction, Gregory assumed that an analytical composition S existed that would satisfy the above Eq. (2.17). Gregory's *reductio* proof then proceeds in two stages. The first one consists in removing the irrationalities contained in Eq. (2.17) by a rational parametric representation in the positive magnitudes a and b.[63] We shall thus have, following Gregory's parametrization,

$$ABP = I_0 = a^2(a + b),$$

$$ABFP = C_0 = b^2(a + b).$$

[62]Cf. *VCHQ*, Prop. XI, pp. 25, 28.

[63]*VCHQ*, Proposition XI, p. 25.

Relying on Eq. (2.17), the terms of the second pair (I_1, C_1) can be parametrized in a and b as well. We have, therefore,

$$ABIP = \sqrt{C_0 I_0}\& = ab(a+b),$$

$$ABDLP = \frac{2C_0 I_0}{I_0 + \sqrt{C_0 I_0}} = 2ab^2.$$

By virtue of the *reductio* assumption, Gregory could also write the identity:

$$S(a^2(a+b); b^2(a+b)) = S(ab(a+b); 2ab^2). \tag{2.18}$$

At this point, Gregory noted that the terms of the left side polynomial, i.e., $S(a^2(a+b); b^2(a+b))$, are inhomogeneous with the terms on the right side, i.e., $S(ab(a+b); 2ab^2)$. In fact, the term a appears, if we expand the expressions that figure as arguments of S, up to the third power, while the right-hand side contains the same terms only up to the second power. The same "imbalance" occurs for the term b. Moreover, Gregory also noted that the right side of the identity contains a monomial term, namely, $2ab^2$, while on the left side, both terms are binomials.

Proceeding based upon these premises, Gregory stated that if the same finite succession S of analytic operations is applied to the expressions $(a^2(a+b); b^2(a+b))$ and $(ab(a+b); 2ab^2)$, then the resulting polynomial on the left side would exhibit either term a or term b raised to a higher power than the corresponding terms of the polynomial on the right side. Along similar lines, Gregory also argued that, under any finite combination of additions, subtractions, multiplications, divisions and root extractions, the polynomial on the left-hand side of the supposed identity will be composed by more terms than the polynomial on the right-hand side.[64] On the strength of these arguments, Gregory concluded that no finite combination S of analytical compositions existed capable of satisfying Eq. (2.18).

But if the above equation could not be satisfied by any analytic composition S, the limit of the convergent series could not possibly be a quantity analytical with the terms of the series either[65]:

> It is evident that the sector $ABIP$ [$A\overset{\frown}{P}B$ in my present account] cannot be composed from the addition, subtraction, division and extraction of roots of the triangle ABP and of the trapezium $ABFP$. We suppose that triangle ABP and trapezium $ABFP$ are quantities analytical with respect to one another; hence the sector $ABIP$ [$A\overset{\frown}{P}B$ in my present account] cannot be analytical with respect to them, that is, it cannot be composed by the addition,

[64] *VCHQ*, pp. 27–28.

[65] See Hofmann (1974, p. 65), Dehn and Hellinger (1943, p. 475) and Lützen (2014, pp. 226–227).

subtraction, multiplication, division and root extraction of the analytical quantities ABP and $ABFP$.[66]

Is Gregory's impossibility argument correct? We could argue that Gregory went as far as to prove that the composition S cannot be formed by a finite sequence of additions, subtractions, multiplications or divisions, and their combinations. In other words, there is no rational function S that solves the functional equation (2.18). However, the matter becomes more complicated when root extraction is evaluated: how could we prove that the left-hand and right-hand terms remain inhomogenous under finitely many root extractions (of any order)?[67] The question does not seem to have been settled to this day, so it is not known whether Gregory's argument may be "repaired" into a cogent proof.[68]

On the other hand, in order to prove that no analytic, or algebraic composition S can exist that satisfies the functional equation (2.17), it is sufficient to show that the tan-function (or the corresponding arctanh-function, in the case of the hyperbola) is not algebraic either.[69] One can be easily convinced of this by considering that the equation: $tan(x) = y$ remains invariant for the transformation $x' = x + \pi$. Therefore, the graph of the tan-function intersects the x-axis at an infinite number of points: this proves that the tan-function has an infinite number of zeroes, hence it is transcendental.

2.4.2 A New Operation

Gregory stressed, in several passages of his book, that, although the ratio between the sector $\overset{\frown}{APB}$ and its inscribed triangle ABP (or circumscribed polygon $ABIP$) could not be expressed by means of analytical quantities, it was not doomed to remain wholly unknown. On one hand, because the sector is the limit of an infinite convergent series of analytical quantities, it can be approximated with ever-increasing precision via analytical quantities. A parallel can be made here with surd numbers like $\sqrt{2}$, which can be approximated by ratios between integers with an ever-increasing degree of precision, even though it can never be exactly expressed by such a ratio.[70]

---

[66] *VCHQ*, p. 29: "Ex hactenus demonstratis manifestum est sectorem $ABIP$ [$\overset{\frown}{APB}$ in my narration] non posse componi ex addictione, subductione, multiplicatione, divisione et radicum extractione trianguli ABP et trapezii $ABFP$. Triangulum ABP et trapezium $ABFP$ supponimus esse quantitates inter se analyticas, et proinde sector $ABIP$ [$\overset{\frown}{APB}$ in my narration] illis analytica esse non potest, hoc est ex quantitatum ipsis ABP $ABFP$ analyticarum additione, subductione, multiplicatione divisione et radicum extractione componi non potest."

[67] Dehn and Hellinger (1939, p. 476).

[68] See Whiteside (1961, pp. 269–270).

[69] Dehn and Hellinger (1943, p. 475).

[70] VCHQ, p. 19: "etiamsi ex praedicto capite non possimus comprehendere rationem inter triangulum ABP et sectorem $ABIP$, possumus tamen eius aliquam habere cognitionem, ex eo quod sector

On the other hand, Gregory conceived of the construction of the infinite convergence series itself as an actual mathematical operation, and a non-analytical one by virtue of the impossibility result proven in the *VCHQ*. This new operation would achieve "the true and legitimate quadrature of the circle in its own kind of proportion." Gregory called this operation a "sixth operation," drawing a further analogy with the operation of root extraction:

> It must be remarked that just as surd numbers are never obtained from the addition, subtraction, multiplication or division of quantities mutually commensurable but only from the extraction of roots, so numbers or non-analytical quantities are never obtained from the addition, subtraction, multiplication, divisions or extraction of root of the analytical quantities, but from this sixth operation, so that our invention adds another operation to arithmetic, and another kind of ratio to geometry.[71]

This parallel rests on the role that impossibility results play, both in geometry and arithmetic. Just like the proof of the irrationality of $\sqrt{2}$ shows that root extraction must necessarily be added to the other arithmetical operations in order to generate surd numbers, so Gregory's proof of impossibility shows that the construction of convergent series must be added to the five arithmetical operations in order to express the area of any central conic sector from the in- and circumscribed polygons. Corresponding to the enrichment of arithmetic with new operations, geometry is enriched with new kinds of ratios (*aliam rationis speciem*). In Gregory's view, the operation of root extraction has a perfect correspondence in geometry in terms of: "the invention of a commensurable proportion which approximates the most closely to an incommensurable analytical proportion," and the sixth operation is represented in geometry via: "the invention of a commensurable proportion which approximates the most closely to our non-analytical proportion."[72] When discussing the operation of root extraction, Gregory might have had in mind methods

---

$ABIP$ est terminatio seriei convergentis datae; et ex hac consideratione possibile est invenire quantitatem datae commensurabilem cuius differentia a sectore $ABIP$ minor fuerit quacumque quantitate proposita ...."

[71] *VCHQ*, p. 5: "Advertendum quoque est sicut numeri fracti nunquam procedunt ex commensurabilium additione, subductione, multiplicatione, divisione, sed tantum ex radicum extractione; ita numeros, vel quantitates non analyticas numquam provenire ex analyticarum additione, subductione, multiplicatione, divisione, radicum extractione, sed ex sexta hac operatione, ita ut haec nostra inventio addat arithmeticae aliam operationem, et geometriae aliam rationis speciem."

[72] Cf. *VCHQ*, p. 5: "Ex illis quinque operationibus arithmeticis, duae sunt tantum simplices, addition & substractio, multiplicatio est composita ex additione, & divisio ex substractione, et extractioradicum, quae in genere nihil aliud est quam inventio proportionis commensurabilis, quae quam proxime accedit ad proportionem analyticam incommensurabilem, componitur ex praecedentis quatuor, & nostra sexta operatio, quae in genere nihil aliud est quam inventio proportionis commensurabilis quam proxime accedentis ad nostram proportionem non analyticam, componitur ex prioribus quinque."

for approximating the result of such an operation by rational estimates.[73] In the same way that Gregory interpreted the operation of root extraction as an approximation of incommensurate quantities by a series of commensurable quantities, Gregory's sixth operation is identified with the construction of a convergent series of analytical quantities or numbers that tend towards an non-analytical quantity or number.[74]

From the impossibility of squaring an arbitrary sector of a central conic by algebraic means (indefinite quadrature), Gregory derived the impossibility of solving the quadrature of the whole circle (definite quadrature). This remarkable conclusion is derived as a mere corollary of his impossibility theorem, as we read at the end of a *Scholium* to proposition 11:

> Since it has been proven that the ratio between the circle and the square on the diameter is not analytical, it will be vain and useless to look for it in the future as it is done now: but leaving analytical quantities aside, I hardly think that there is a quantity more known than these terminations of our convergent series, as will very clearly appear from the next propositions.[75]

Gregory states here that the ratio between the circle and the square built on its diameter could not be found using the five analytical operations alone. As the second part of the book shows in more detail, the area of the circle (as well as any given sector of the ellipse or the hyperbola) must be expressed as the limit of an infinite convergent series, whose truncations yield better and better numerical approximations to its real non-analytical measurement.

The impossibility of solving the definite squaring of the circle analytically was so downplayed in the *VCHQ* that Wallis, upon reading the book, even doubted whether Gregory had, in fact, made this claim and intended to prove the impossibility of squaring the circle analytically (cf. Wallis 2012, p. 28).

Doubts are dispelled, however, by considering Gregory's correspondence. For example, an interesting discussion can be found in a letter published in the *Philosophical Transactions of the Royal Society*, which Gregory addressed to the latter on December 25, 1668, also stimulated by the criticisms advanced by Huygens in the aftermath of the publication

[73]For the square root, an example of such a method was offered by the series that Gregory had discovered and illustrated in a letter to Huygens, cited above, in Sect. 2.3.5. This example perfectly illustrated, using only the mathematical tools presented in the *VCHQ*, how the surd number $\sqrt{z}$ (with z positive integer) could be expressed through an infinite double convergent sequence of rational numbers.

[74]*VCHQ*, p. 29.

[75]*VCHQ*, p. 29: "cum enim demonstratum sit rationem circle ad diametri quadratum non esse analytica, vanum certe erit et ineptum illam sicut talem imposterum quaerere: at reiectis quantitatibus analyticis, vix credo ullam posse esse notiorem hisce nostrarum serierum convergentium terminationibus, sicut ex sequentis plenissime apparebit."

of the *VCHQ*.[76] In the letter, Gregory gave a more extensive proof for the impossibility of expressing the ratio between a circle and the square on its diameter analytically, which can be briefly synopsized:

Corollary 2.1 *In a given circle, the ratio between the area of the circle and the square of the diameter is not a magnitude analytical with the radius.*

Proof For the sake of contradiction, let us suppose that the ratio between the area of the circle, namely, c, and the area of the square, namely, q, is a quantity analytical with respect to the radius r. Hence, c will be analytical with q. But the ratio between the circumscribed square q and the inscribed square q' is also analytical with respect to the radius, thus there is an analytical composition f, such that $c = f(q, q')$. Since we have assumed that the circle can be analytically composed from the in- and circumscribed squares, then the quarter of the circle can be analytically composed from the inscribed triangle with area $\frac{r^2}{2}$ and the circumscribed square with area r^2. But this contradicts Proposition XI of the *VCHQ*, which proves that the area of a sector cannot be composed analytically from its in- and circumscribed polygons, if these polygons are mutually analytical.

As we shall see in the next sections, Gregory's claim to have solved the definite quadrature of the circle was to be vehemently denied by Huygens, who did not find the above argument satisfying. Later on, a similar criticism was to be reenacted and expanded by Leibniz.

Before taking into consideration Huygens' criticism, let us remark that there is an obvious objection to be raised against the impossibility of inferring the definite quadrature of the circle from Proposition XI. This objection may go like this. Since there are infinitely many sectors in a circle, some of these could have areas analytical or even commensurable with respect to the area of their inscribed triangle I_0. Thus, Gregory's impossibility argument does not hold for any and every sector. In fact, there is a simple answer to this objection: Gregory's theorem, formulated in Proposition XI, does indeed not hold for any and every sector without distinction, but rather only for such sectors whose in- and circumscribed polygons are analytical one with respect to the other, or have an analytical

[76] The letter appeared in February 1669, and can be found in Huygens (1888–1950, Vol. 6, p. 309).

ratio. Gregory made this point in an interesting *Scholium* to proposition XI, in where he noted:

> But someone may by chance say that the ratio between the triangle APB and the sector $ABIP$ [$A\overset{\frown}{P}B$ in our account] can vary in every respect; hence these figures can be in any given ratio one to the other, either analytical or commensurable. I answer that this is very true, but in that case the ratio between the triangle ABP and the trapezium $ABFP$ will not be analytical.[77]

Thus, if the triangle ABP is analytical with respect to the sector $A\overset{\frown}{P}B$, then the polygons ABP and $ABFP$ are not analytical with respect to one another. This follows logically from Gregory's Proposition XI, and implies that the radius and the chord of the sector are not analytical with respect to one another. Therefore, it is true that the areas of certain sectors are analytical with respect to the area of their corresponding inscribed triangle ABP, but, in these cases, it will be impossible to compute the area of the circumscribed polygon $ABIP$ analytically from the triangle and the sector.

2.5 Reception and Criticism of Gregory's Impossibility Argument

2.5.1 Huygens' First Objections

As mentioned in Sect. 2.2, a controversy between James Gregory and Christiaan Huygens began in 1668, almost 1 year after the publication of the *VCHQ*. The controversy originated from Huygens' first account of the *VCHQ*, which appeared in the *Journal des Sçavans*, in July 1668, and brought Gregory's impossibility claim under criticism.

In this section, I shall explore Huygens' objections contained in the July address, as well as the initial reactions of Gregory and other mathematicians. In particular, I shall emphasize the role of John Wallis, whose involvement in the controversy was solicited by his colleagues in the Royal Society.

I shall build my survey on valuable previous work, in particular, the classical articles contained in the *Tercentennial Memorial Volume*, edited by Turnbull (see Turnbull 1939), such as Dehn and Hellinger (1939) and Dijksterhuis (1939), and the more recent (Lützen 2014).

[77]*Scholium*, prop. XI, p. 19: "Sed dicet forte aliquis rationem inter triangulum APB et sectorem $ABIP$ omnifariam variari posse; et proinde posse esse inter se in ratione qualibet data, sive analytica sive etiam commensurabili: respondeo hoc esse verissimum, sed in hoc casu ratio inter triangulum ABP et trapezium $ABFP$ non erit analytica."

Huygens' review from July 1668 contains several objections against the validity of Gregory's impossibility proof.[78]

The first two objections are of little relevance, as they stem either from Huygens' confusion about what it means for a composition S to be invariant for the terms $(a^3 + a^2b; ab^2 + b^3)$ and $(a^2b + b^2a; 2b^2a)$,[79] or from a mistake effectively committed by Gregory, but which had no serious consequences on the rest of his argument.[80] In response, Gregory admitted his mistake and replaced the faulty series with a correct one, namely, the geometrical series explored in Sect. 2.3.5 above.

Contrastingly, Gregory would find it more difficult to reply to the third criticism raised by Huygens:

> Even if this is true [namely, that the "terminatio" is analytical with respect to the double sequence] when the limit is found by the method taught by Gregory, we cannot derive a general conclusion, unless we suppose that we can find only by his method the limit of a sequence of magnitudes, that he calls convergent, or that, if we find them by another route, we will be able to find it by his method.[81]

With the above objection, Huygens made explicit what I have called above a "tacit assumption" grounding Gregory's technique for computing the limit of a convergent series, namely: if a convergent series tends to an analytical limit, then this limit can be found according to the method described in the *VCHQ*. This assumption could not be taken for granted, Huygens correctly pointed out, unless one advanced the less-than-obvious supposition that the limit of a sequence of magnitudes could be found only by Gregory's method, or by an equivalent method.

Gregory promptly dismissed Huygens' criticism. In his first reply from July 13, 1668, Gregory had already remarked that Huygens had not offered any reason that might give a solid ground for doubting (*solidam dubitandi ratione*) the validity of Proposition XI and its *Scholium*. For instance—Gregory continued, passing now to the counterattack—Huygens never provided a counterexample, namely, a convergent series of analytical terms, whose limit would be analytical with respect to the terms yet not computable through Gregory's method.[82]

---

[78]Huygens' report can be found in Huygens (1888–1950, Vol. 6, pp. 228ff.). Gregory's reactions to Huygens' first piece appeared in July 1668 (Gregory 1668a, reprinted in Huygens 1888–1950, Vol 6, p. 240).

[79]Huygens (1888–1950, Vol. 6, pp. 242–243). See also Scriba (1957, pp. 18–19).

[80]Cf. Huygens (1888–1950, Vol. 6, p. 230).

[81]Huygens (1888–1950, Vol. 6, p. 229). See also Dijksterhuis (1939, p. 483).

[82]As Gregory protested: "I would certainly like that this very Noble Man will assign me a convergent sequence that, together with its limit, will refute our corollary or, if he cannot find it, I wish only a solid ground for doubting." See Huygens (1888–1950, Vol. 6, p. 240). The corollary (*consectarium*) to which Gregory referred can be found at the end of Proposition X, *VCHQ*, p. 24: "et proinde ad inveniendam cuiuscumque seriei convergentis terminationem; opus est solummodo invenire

In the same letter, Gregory reinforced this generic criticism with an attempt to prove the assumption that sounded so problematic to Huygens, and that we can phrase as such:

Theorem 2.5 *If, for every sector z, the area of a central conic sector is analytical with respect to the terms of the convergent series, then it can be computed using the method expounded in Propositions VII-X of the* VCHQ.

A correct proof of this implication would have repudiated Huygens' criticism. Unfortunately, Gregory did not manage to offer a cogent proof, and his argument remained, in the end, unconvincing.

Gregory's idea is to prove that if the area of the sector is analytical with respect to the convergent series of in- and circumscribed polygons, then an analytical composition invariant with respect to any pair (I_n, C_n) exists. Let us assume, for the sake of contradiction, that no such analytical composition exists. Since the sector, let us call it Z, is, by hypothesis, analytical with respect to any pair (I_n, C_n), there exists an analytical composition S such that: $S(I_0, C_0) = Z$. However, by the *reductio* assumption, the same composition applied to the pair (I_1, C_1) might not yield the same quantity Z. Let us suppose that this is indeed the case and $S(I_1, C_1) = K$, with $K \neq Z$. Let the difference $|K - Z|$ be equal to a quantity α. By the convergence of the series, there exists an n for which $|I_n - C_n| < \alpha$. Moreover, since Z expresses the area of the sector, the following inequality also holds: $I_n < Z < C_n$ (with respect to the case of the circle and the ellipse, while the inequalities must be inverted in the case of the hyperbola) in virtue of the definition of a convergent series. Finally, K being formed by the pair (I_1, C_1) in the same way that Z is composed by (I_0, C_0), the following inequality is obtained for an elliptical or circular sector: $I_{n+1} < K < C_{n+1}$, which involves: $I_n < K < C_n$. From the previous pair of inequalities, we conclude that the difference between Z and K must also be a quantity greater than I_n and smaller than C_n, thus: $|K - Z| < \alpha$, which contradicts our assumption that $|K - Z| = \alpha$. From this contradiction, it follows that $Z = K$, hence there exists an analytical composition S, invariant with respect to (I_0, C_0) and (I_1, C_1).

However, Gregory failed to prove a crucial result, because he took for granted the following implication: if $I_n < Z < C_n$, then $I_{n+1} < K < C_{n+1}$. This is not true in general for convergent series, as a simple counterexample found in Lützen (2014, p. 229) proves. Therefore, as it stands, Gregory's proof of Theorem 2.5 could hold for certain convergent series, but does not hold universally. Moreover, it remains unclear as to whether it holds for the series of polygons converging to any given sector of the circle.

---

quantitatem eodem modo compositam ex terminis convergentibus primis, quo componitur eadem quantitas ex terminis convergentibus secundis" ("And, conclusively, in order to find the *terminatio* of any convergent sequence, one only needs to find a quantity composed in the same way from the first convergent terms, as from the second convergent terms.").

Without providing any counterexample, Huygens rejected Gregory's argument, and remained of the opinion that the impossibility of the analytical quadrature of the central conic sectors was far from having been proven. Huygens expressed his utter dissatisfaction in a letter to the *Journal des Sçavans* from November 1668, which also constitutes the second and final public contribution by Huygens to the controversy.

In a series of private notes, probably written as a draft for the November letter, Huygens made his position more precise:

> Although Gregory, in the answers given to my objections, has supplemented some defects in his proofs, he will let me tell that, after this, a lot is needed to prove the impossibilty of the quadrature of the circle well ... Even after the proof he [Gregory] has given to supplement his Proposition 11, we can only conclude that not all the sectors of the circle are in an analytical ratio with their inscribed or circumscribed figures. Which is a wholly different thing than to say that no sector of the circle is. He who says *non omnis* does not say *nullus*. And thus it is not sufficient to demonstrate that the sector of the circle is not analytical *indefinite* to its inscribed figure, but it has to be proven that the same result holds *in casu omni definito*.[83]

In the same manuscript, Huygens detailed the reason why Proposition XI of the *VCHQ* failed to convince him of the impossibility of squaring the circle "in every definite case":

> Whenever between the quantities a and b of his Proposition 11 there is a numerical proportion, we will not be able to state how another proposed number is composed from the first or the second terms of the convergent series. Consequently, in all these cases Proposition 11 does not prove that it is impossible to find the termination by any other method than the one given by the author.[84]

Following Huygens' suggestion, let us consider a concrete example, namely, the sector equal to one third of the circle. In such a case, the first inscribed triangle (a) is equal to one third of the first circumscribed polygon ($3a$). What conclusion are we entitled to draw regarding the relation between the converging polygonal series and the area of the sector?

[83]Huygens (1888–1950, Vol. 20, p. 308): "Quoy que M. Gregorius dans la response qu'il a faite a mes objections ait supplee quelques defauts qu'il y avoit dans ses demonstrations, il me permettra de dire qu'il s'en faut tant qu'apres cela l'impossibilité de la quadrature du cercle soit bien prouvee ... Car mesme apres la demonstration qu'il a donnee pour supplement de sa propos. 11. qu'est ce qu'il en peut conclure, si non que tout secteur de cercle n'est pas en raison analytique a sa figure rectiligne inscrite ou circonscrite. Ce qui est tout autre chose que de dire que nul secteur de cercle ne l'est. Qui dit *non omnis* ne dit pas *nullus*. Et ainsi il ne suffit pas de demonstrer que le secteur de cercle a sa figure inscrite n'est pas analytique indefinite, mais il faut demonstrer que c'est la mesme chose in casu omni definito."

[84]Huygens (1888–1950, Vol. 20, p. 308): "Depuis qu'entre les quantitez a et b dans sa prop. 11 il y aura proportion numerique, l'on ne pourra plus dire de quelle façon un autre nombre proposè est composè des premiers ny des seconds termes de la suite convergente. Et par consequent en tous ces cas la propos. 11 ne demonstre point qu'il soit impossible de trouuer la terminaison par quelque autre methode que celle de l'autheur."

Huygens has no doubt: there is no way to claim that the sector corresponding to one third of the circle is not analytical with its corresponding inscribed triangle.

These very arguments are summarized in the second letter published in November 1668. At that point, Huygens could have realized that Gregory's Theorem 2.5 lacked generality, since, as we know, it is valid for some convergent series, but not for all.

If we venture the hypothesis that Huygens was aware of this flaw, despite never mentioning it in his notes and correspondence, then Huygens' answers make full sense. Indeed, even after Gregory's proof of Proposition 11, one may not draw any conclusions about the impossibility of squaring the sector equal to one third of the circle, or any other special sector. To state the impossibility of squaring a specific sector of the circle, Huygens retorts, one would have to show:

> ... not only that the sector of the circle is not analytic *indefinite* to its inscribed figure, although this proof still has a certain beauty; but that this is also true in any definite case.[85]

It goes without saying that this is a formidable task, and not one Gregory could have easily achieved. Huygens was thus confident that: "it still remains uncertain whether the circle and the square on its diameter are not commensurable, that is, whether they have the ratio of a number to another number."[86] By consequence, Huygens kept the problems of the impossibility of the "indefinite" quadrature of the circle, i.e., the problem of proving that there is *at least one* sector whose area is not analytical with the converging series of polygons, distinct from the one of the impossibility of the "definite" quadrature, i.e., the problem of proving that there is *no* sector that is analytically composed from the converging series of polygons. Gregory's Proposition XI could, at most, prove the former, but in no conceivable way could it be adjusted to prove the latter.

In Huygens' letter from November 1668, we find another objection against Gregory's conclusion regarding the impossibility of the definite quadrature of the circle. This objection builds on the assumption that there exists no analytical composition f such that the area of a sector is an explicit function of its in- and circumscribed polygons. Even admitting this, it does not follow that it is impossible to express the area of a sector as an implicit function of the polygonal areas. For example, it does not follow that the area of the circle cannot satisfy an equation of a degree higher than the fourth. In such cases, Huygens noted, it is not known how the root can be expressed as an explicit function of its coefficients (as we now know, it is indeed impossible), although it is known how it can be constructed geometrically, using the Cartesian technique of the construction of equations.

[85]Huygens (1888–1950, Vol. 6, p. 273): "Pour conclure donc que la raison du Cercle au Quarré de son diametre n'est pas analytique, il falloit demontrer non seulement que le Secteur de Cercle n'est pas analytique indefinite à sa figure inscrite, quoyque cette demonstration ne laisse pas d'avoir sa beautè; mais que cela est vray aussi in omni casu definito ... "

[86]Huygens (1888–1950, Vol. 6, p. 273): "Il demeure encore incertain si le Cercle et le Quarré de son diametre ne sont pas commensurables, c'est à dire à raison de nombre à nombre."

Thus, the area of the sectors could be analytical with respect to its in- and circumscribed polygons I_0 and C_0, even if there exists no explicit analytic composition, or function f such that $A = f(I_0, C_0)$.[87]

2.5.2 Wallis Against Gregory's Impossibility Theorem

Huygens' objections were endorsed by Wallis, who had a brief exchange of letters with the former, and soon after sent to Brouncker a review of the *VCHQ*, read at a meeting of the Royal Society on December 3/13, 1668 (see Wallis 2012, p. 25).

In that letter, Wallis set out to analyze Gregory's arguments so as to present an "anatomy"—so he says—of the first eleven propositions and the deductive structure of the *VCHQ*. He thus concluded that Gregory had proven the following in Proposition XI of the *VCHQ*:

23. The *Sector indefinitely taken* is not analyticall with its Triangle & Trapezium. For the *Sector indefinitely taken* is this termination.
24. Therefore not *every sector*. For if every sector, then any sector whatsoever; that is, the sector indefinitely taken.
25. Therefore *no* sector: At least no sector which is analyticall with the whole Circle, and whose Triangle & Trapezium are analyticall with the square of the Radius or of the Diameter. Which consequence, I doubt, will hardly be made good.
26. Therefore the Circle cannot be by that process analytically squared. For that process supposeth some such Sector which shall be analyticall with the Circle, et with its own Triangle & Trapezium.[88]

Along with Huygens, Wallis contested the validity of the inference from n. 23 above (the impossibility of the indefinite quadrature of the circle) to 26 (the impossibility of the definite quadrature of the circle), and therefore the validity of Gregory's claim to have proven the impossibility of the definite quadrature of the circle (specifically 25 and 26).

According to Wallis, the logical weakness in Gregory's demonstration could be summed up in the following, flawed syllogism:

His argument seemes to mee [sic] to be ⟨*are*⟩ such a forme as this; If the Sector be so compounded, the circle may be analytically squared; But the Sector is not so compounded; Therefore the circle cannot be analytica[lly] square[d.] Which Syllogism is peccant in form,

[87]Huygens (1888–1950, Vol 6, p. 273): "Je dis de plus, que les quantitez a et b demeurant indeterminées, la terminaison se reduira peut-estre à quelque equation de celles dont on ne peut pas donner la racine; sans que le contraire se puisse prouver par sa Proposition XI ny par son supplement: Et neantmoins si cette terminaison estoit reduite à quelque equation de cette nature, je croirois que la Quadrature seroit trouvée geometriquement; et le Probleme se pourroit resoudre par l'intersection de quelques lignes courbes qu'on reçoit en Geometrie."

[88]Wallis (2012, p. 35).

> though the propositions be true; & therefore the conclusion follows not. Unlesse we suppose not onely the Consequence of the Major Proposition to be demonstrated; but the Converse of that consequence; which is not done.[89]

Wallis acknowledged that such a syllogism was "peccant in form," so that the conclusion, even if true, did not follow from its premises. This is the same problem uncovered by Huygens in his first reply to Gregory, but Wallis passed it over for a new objection and used it as a pretext to advertise his own mathematical contributions. In fact, he then proposed to amend Gregory's proof starting from the very impossibility of expressing the ratio between the area of the whole circle and its in- or circumscribed squares as an analytical quantity, and then deriving, by contraposition, Proposition XI of the *VCHQ*. Wallis' starting point would be his own treatise *Arithmetica Infinitorum* (1656), which contains an argument suggesting that the ratio between the area of the circle and in- or circumscribed square cannot be expressed as a rational or surd number (Proposition CXC).

This argument actually reduces to a passing remark:

> And indeed I am inclined to believe (what from the beginning I suspected) that this ratio we seek is such that it cannot be forced out in numbers according to any method of notation so far accepted, not even by surds (...) so that it seems necessary to introduce another method of explaining a ratio of this kind than by true numbers or even by the accepted means of surds.[90]

Wallis did not pursue this argument further, hence it is at least dubious as to whether he had managed to give a fully-fledged proof of the impossibility of squaring the circle that was any better than that which Gregory had offered.

Less specious and more interesting is another argument proposed by Wallis, concerning an analogy between the case of the quadrature of the circle and the problem of trisecting the arc:

> In this Equation for the Trisection of an Arch $3r^2a - a^3 = r^2c$ indefinitely taken; the Root a, is not Analyticall with r and c, (that is, the Proportion of the Chord of the Single Arch, to that of the Triple arch & the Radius, cannot be universally designed by those he calls Analyticall operations; or the value of a analytically compounded of r and c, as he speakes; that is, it cannot be designed by commensurable numbers & surd Roots:) as Chartes, Schoten & others agree. Yet in some cases (though not universally) it may happen to be not onely Analytical, but even commensurable (as, for instance, if $c = 2r$ be the Subtense of a Semicircle, a may be equal to r or to $2r$).[91]

[89]Wallis (2012, p. 35).

[90]Wallis (2004, p. 161). See also Lützen (2014, pp. 218–219) and Panza (2005, pp. 77–78).

[91]Wallis (2012, p. 30).

In Book III of *La Géométrie*, Descartes indeed gave instruction as to how to reduce this problem of angular trisection to the solution of a cubic equation of the form $a^3 = 3r^2a - r^2c$, where r is the radius of a given circle, c the chord subtending an arc of the given circle, and a the chord subtending the third part of c.[92]

In light of the above formula, one might think that the relation between the chord c subtending a given arc and the chord subtending its third is given by a cubic equation, and is thus analytical. But a problem arises when one tries to solve the equation by radicals, and thus express a as an explicit function of c. In fact, the cubic equation connected with the problem of trisecting an angle is an instance of the so-called *casus irreducibilis*, namely, a third degree equation involving square roots of negative quantities, as shown by the Cardano rule[93]:

$$a = \sqrt[3]{\frac{r^2c}{2} + \sqrt{(\frac{r^2c}{2})^2 - (\frac{3r^2}{3})^3}} + \sqrt[3]{\frac{r^2c}{2} - \sqrt{(\frac{r^2c}{2})^2 - (\frac{3r^2}{3})^3}}.$$

One can geometrically verify that the term under the square root: $(\frac{r^2c}{2})^2 - (\frac{3r^2}{3})^3$, is negative, because any chord in the circle different from the diameter is shorter than the diameter. However, for Descartes, the emergence of imaginary or "impossible" quantities in the expression of a was not a threat to the very existence of the roots of the cubic, which received a clear geometrical interpretation within the context of the trisection problem. In fact, the roots of the cubic equation expressed the lengths of certain chords within a given circle with radius r, or it expressed the sides of certain triangles determined by the equation itself.[94] Descartes even added, in a rather marginal remark, an interesting consideration: the relation between the unknown (geometrically interpreted, this is the chord subtending the third of a given angle) and the known terms (the chord subtending the given arc and the radius of the circle) might be expressed by means of an equation involving the use of a "new" symbol that represents the geometrical operation of taking the chord subtending an arc equal to a third of a given arc, in the same way as: "...the symbol $\sqrt{3}$ is used to represent the side of a cube."[95]

Probably because he lacked a mathematical apparatus for dealing with transcendental relations, Descartes did not develop this suggestion any further, nor did he explain the nature of such a relation. As far as I could ascertain, Wallis relied on Descartes' opinion too. However, using trigonometric functions, we can nowadays easily express the nature of such a relation. The fundamental idea boils downs to the triple angle formula, namely, the

[92]See, for instance, Descartes (1897–1913, Vol. 6, p. 470): "La façon de diviser un angle en trois." The problem is also discussed in Descartes (1659, Vol. I, p. 345).

[93]Cf. Descartes (1897–1913, Vol. 6, pp. 471–472).

[94]Descartes (1897–1913, Vol. 6, p. 474).

[95]Descartes (1897–1913, Vol. 6, pp. 474–475). See also Bos (2001, p. 379).

equation that connects the cosine of an angle (θ) to the cosine of its triple (3θ): $cos3\theta = 4cos^3\theta - 3cos\theta$.

In order to express the chord of a given arc as an explicit function of the chord of its triple, it is sufficient to make the cubic equation $a^3 = 3r^2a - r^2c$ look like the equation: $cos3\theta = 4cos^3\theta - 3cos\theta$. Let us first transform, for simplicity's sake, our original cubic equation into the reduced cubic form: $y^3 + py + q = 0$ ($a = y$, $-3r^2 = p$, $r^2c = q$). We set: $y = ucos\theta$ (the nature of the parameter u will soon be clear), so that the equation $y^3 + py + q = 0$ becomes

$$(ucos\theta)^3 + pucos\theta + q = 0. \tag{2.19}$$

Next, we multiply all of the members of the above Eq. (2.19) by the factor $\frac{4}{u^3}$, and we obtain

$$4cos^3\theta + \frac{4p}{u^2}cos\theta + \frac{4q}{u^3}.$$

The new equation can be compared termwise with the triple angle formula $cos3\theta = 4cos^3\theta - 3cos\theta$. Thus, we observe that: $\frac{4p}{u^2}cos\theta = -3cos\theta$, hence $u = \sqrt{\frac{-4}{3}p}$ (since $p < 0$, u is well defined). Thus, we shall also have: $-cos3\theta = \frac{4q}{u^3}$, and, substituting the value $\sqrt{\frac{-4}{3}p}$ into u, we finally obtain

$$-cos3\theta = \frac{3}{2}\frac{q}{p}\sqrt{\frac{-3}{p}}.$$

Considering that we have set: $y = ucos\theta$, the roots of the cubic equation can be expressed as explicit functions of the parameters p and q as follows:

$$y_1 = \sqrt{\frac{-4}{3}p}cos(\frac{1}{3}arccos(-\frac{3}{2}\frac{q}{p}\sqrt{\frac{-3}{p}})),$$

$$y_2 = \sqrt{\frac{-4}{3}p}cos(\frac{1}{3}arccos(-\frac{3}{2}\frac{q}{p}\sqrt{\frac{-3}{p}}) + \frac{2}{3}\pi),$$

$$y_3 = \sqrt{\frac{-4}{3}p}cos(\frac{1}{3}arccos(-\frac{3}{2}\frac{q}{p}\sqrt{\frac{-3}{p}}) + \frac{4}{3}\pi).$$

Substituting, in the above solution, the terms corresponding to the radius (r), to the chord (c) of the given arc and to the chord of the third of the arc (a), we arrive at three solutions:

$$a = 2rcos(\frac{1}{3}arccos(-\frac{c}{2r})), \tag{2.20}$$

$$a = 2rcos(\frac{1}{3}arccos(-\frac{c}{2r}) + \frac{2}{3}\pi), \tag{2.21}$$

$$a = 2rcos(\frac{1}{3}arccos(-\frac{c}{2r}) + \frac{4}{3}\pi). \tag{2.22}$$

Wallis might not have had a definite argument to establish that the trigonometric relation between a and c is generally expressed by a non-analytical composition (and, in fact, I have not found any such argument), but he could well have suspected, as Descartes might also have, that this operation was not reducible to the five arithmetical operations. On the other hand, Descartes and Wallis obviously knew that there existed at least one arc for which the relation between the subtended chord and the chord subtending its third part is analytical (Wallis 2012, p. 30). This is the case for the semicircumference, for instance, whose chord is equal to the diameter itself (i.e., $c = 2r$). In fact, substituting $c = 2r$ in Eq. (2.20) etc., we obtain

$$a = 2rcos(\frac{1}{3}arccos(-1)), \tag{2.23}$$

$$a = 2rcos(\frac{1}{3}arccos(-1) + \frac{2}{3}\pi), \tag{2.24}$$

$$a = 2rcos(\frac{1}{3}arccos(-1) + \frac{4}{3}\pi). \tag{2.25}$$

The second case, where $a = -2r$, does not correspond to a real geometrical situation, while in the first and third cases, we have that $a = r$. This solution concords with Wallis's remark above.

The case of the trisection of an arc thus offered Wallis motivation to doubt the legitimacy of Gregory's inference of the impossibility of the definite quadrature of the circle from the impossibility of the indefinite one. By analogy with the case of the diameter and the chord subtending the semi-circumference and the angle of 60°, one might suppose that the area of the whole circle or of one of its analytical submultiples could be expressed as an analytic or commensurable function of the chord and the radius, even if the indefinite analytical quadrature of the circle was proven to be impossible.[96]

In conclusion, behind Huygens' and Wallis's objections, there seems to be a disagreement with Gregory on the meaning of the "indefinite" quadrature problem and its impossibility. Whereas Gregory believed himself to have proven that there exists no analytical composition for a sector such that the area of the sector can be expressed through

[96]Wallis (2012, p. 30).

said composition in terms of its in- and circumscribed polygons, or in terms of their chord and radius analytical one to the other, Huygens read Gregory's theorem as proving a much less general result, namely, that there exists (at least) one sector that is non-analytical with its converging series of polygons. Wallis sided with Huygens' objection, but based his criticism, as we have seen, on an interesting analogous case of the trisection of an arc.

It seems to me that, by stressing this analogy, Wallis came close to understanding the problem of the "indefinite" quadrature of the circle as the problem of finding a (transcendental) relation between the first pair of polygons and the area of a sector or, using trigonometry, between the radius, the chord of a given sector and the arc delimiting the sector itself. In the case of the trisection, although there exists no analytical function relating the chord subtending a given arc to the chord subtending its third, there are particular cases or arcs whose subtending chord is analytical with the chord subtending the third of the arc. Likewise, why should one exclude the possibility that, although the relation between a chord and the arc of a circle is generally non-analytical, the chord and the arc could not be analytical for special sectors?[97] This relation being a trigonometric relation too, we realize that these examples are strikingly similar. However, we also know that there exists no special squarable sector of the circle (or special rectifiable arc) whose chords and radius are mutually analytical. Yet, such knowledge goes well beyond the mathematical developments in the Seventeenth Century.[98]

2.5.3 Gregory's Last Reply

Gregory replied to Huygens' and Wallis's joint attack in a letter to Oldenburg, published in February 1669 in the *Philosophical transactions of the Royal Society*.[99] This was also the closing act of the controversy between Gregory and Huygens.

In his reply, Gregory rejected the objections advanced in Huygens' second letter to the *Journal des Sçavans*, and claimed once more that the impossibility of the definite quadrature of the circle could and should be deduced from the impossibility of the indefinite one, which he had proven in the *VCHQ*.

Firstly, he defended this viewpoint on the basis of the following methodological remark:

> Indeed, I wish that this Noble Man considered that any full solution of a problem is indefinite. In fact particular methods, being infinitely many, cannot all be exhibited, nor can they be determined by the formulation of the problem common to all of them. For this reason we need a general or indefinite method regulative of the particular ones. I certainly acknowledge

[97] See also Lützen (2014, p. 232).

[98] I quote here from Lützen (2014): "it follows from Lindemann-Hermites theorem that v and $sin v$ cannot be algebraic simultaneously. Thus constructible points on the unit circle correspond to transcendental arcs and areas." (Lützen 2014, p. 231).

[99] Gregory (1668c), reprinted in Huygens (1888–1950, Vol. 6, pp. 306–311).

that particular methods of solution are often found by chance, without the aid of the general one, however geometers must recognise that there is or can be no particular method that is not, in the last analysis, the expression of some indefinite method.[100]

Apparently, Gregory had not moved from his initial positions. In contrast to his adversaries, he did not conceive the problems of the indefinite and the definite quadrature of the circle as distinct problems. Thus, if the whole circle could be analytically squared, then there existed a "general" or "indefinite" method for squaring any sector of the circle from which the solution of the circle-squaring problem could be deduced. Conversely, if no general methods for solving the indefinite quadrature of the circle by analytic means existed, no particular analytical method for solving the definite quadrature of the circle existed either.[101]

In order to defend his position and "leave no room for sophistries,"[102] Gregory offered a final argument. This new argument proceeds by *reductio*.[103] Let us consider ϕ, a definite sector of the circle, such as its quadrant. Applying Gregory's construction, we can form a convergent series of in- and circumscribed polygons, starting from an inscribed triangle (let us call it I_0) and a circumscribed square, whose area is the double of I_0 (thus, we shall call it $C_0 = 2I_0$). Let us assume, by *reductio* hypothesis, that the area of the circle is analytical with respect to the square on its diameter, from which it immediately follows that ϕ is analytical with respect to I_0 and $2I_0$. Thus, there exists an analytic composition S such that $S(I_0, 2I_0) = S(\phi) = K$, where the quantity K is analytic with respect to I_0 and C_0. On the other hand, if ϕ is an analytic composite of the initial polygons I_0 and C_0, then it will be an analytic composite of the other pairs of polygons in the convergent series, and we shall have: $S(I_0, C_0) = S(I_1, C_1) = \ldots = S(I_n, C_n) = \ldots S(\phi) = K$. Since the expression $S(I_0, 2I_0) = K$ is a finite polynomial equation, it can be written in Cartesian terms as $S(a, x) = K$, where the unknown corresponds to the sector $C_0 = 2I_0$. However, since S has been supposed invariant with respect to the pairs (I_n, C_n) (for any n), the same equation will hold for any such pair, and x will take infinitely many values (namely, for any n, $x = C_n$). Therefore, Gregory concluded, the equation: $S(x, a) = K$ cannot have

[100]Huygens (1888–1950, Vol. 6, pp. 307–308): "Velim enim Nobilissimum Virum considerare, Omnem plenam Problematis solutionem esse Indefinitam. Nam methodi Particulares, cum sint Jnfinitae, exhiberi omnes nequeunt; neque dirigi possunt tenore Problematis, quippe illis omnibus communi: Jdeoque requiritur methodus Generalis seu Indefinita, Particularium directrix. Agnosco utique methodos Particulares casu saepe inveniri absque ope Generalis, attamen fatendum est Geometris, nullam esse nec posse fieri Methodum Particularem, in quam resolubilis non sit methodus Indefinita."

[101]"Si igitur methodus Indefinita omni resolutioni sit impervia (ut in Propositione 11ma est demonstratum) eodem modo omnes Particulares resolutionem etiam respuent; proindeque tam Definita quam Indefinita nullam compositionem agnoscit."

[102]Huygens (1888–1950, vol. 6, p. 308): "Ne tamen ullus reliquatur cavillationi locus, undecimam nostram *Propositionem* etiam *in definitis* hic demonstrabimus."

[103]Cf. Dehn and Hellinger (1939, p. 476).

a fixed finite degree. Thus, a contradiction, because a finite polynomial equation cannot have infinitely many roots.

As such, the *reductio* argument concocted in Gregory's letter is a *non-sequitur*, as can be quickly ascertained. In fact, if, as Gregory assumed, the argument holds for any convergent series, it will lead to the absurd conclusion that no series exists whose limits are analytical with respect to their terms. This is obviously false, as the series converging to the geometrical mean of the initial terms, discussed by Gregory himself in July 1668, shows.

In the end, the remarks of Gregory's that we have just reviewed failed to satisfy his adversaries. They perceived the obstinate defense of his original positions as a stubborn effort whose sole effect was that of entangling the discussion and estranging Gregory from the support of his most influential colleagues.[104]

Moreover, by the end of 1668, the controversy had taken on harsh tones, which pushed the members of the Royal Society to seek its closure. By the beginning of 1669, the Society had publicly distanced itself from the dispute, by publishing Gregory's letter of February 14, 1669, as his final reply, "but withal, that care should be had of omitting all that might be offensive".[105] The policy of the Society was clearly expressed by the words of its secretary Henry Oldenburg, who, in a letter to Huygens from February 1669, explained that the publication of Gregory's latest reply would also mark the final act of the controversy:

> And as for Mr. Gregory's reply to your answer, it will be inserted, with all respect due to you, in the Transactions of this month. However, I would strongly desire that this dispute will be ended for good, and that we would not need to entertain the readers of these Journals with such details, which express excitement and animosity between eminent persons, and in this way the idea that mathematical knowledge is uncertain.[106]

[104]We can glean this conclusion from Huygens' remarks in a letter to Oldenburg, from March 30, 1669: "Il me semble par la response de Monsieur Gregory qu'il s'est trouvé fort embarassé de mes derniers instances, car au lieu d'y respondre pertinemment, il ne cherche qu'a embrouiller tellement la dispute, et la rendre si obscure que personne n'y comprendra dorenavant rien" (in Huygens 1888–1950, Vol. 6, p. 391). "La response de M. Gregory" to which Huygens refers is obviously the last letter published in the *Transactions* in February 1669 (Gregory 1668c). Again, Huygens wrote in similar tones to Robert Moray, in March 1669: "la derniere response de Monsieur Gregory sur le suject de la Quadrature, ou il n'a rien fait qui vaille, et je voudrois bien scavoir s'il y a aucun des geometres par de la qui prenne pour des demonstrations ce qu'il donne pour telles. J'ay de la peine a m'imaginer qu'il le croye luy mesme, et il me paroit plus vraysemblable qu'il s'est voulu sauver dans l'embaras et dans l'obscurité" (in Huygens 1888–1950, Vol. 6, p. 396).

[105]See Gregory (1668c).

[106]Huygens (1888–1950, Vol. 6, p. 358): " Et quant à la replique de Monsieur Gregory à vostre responce, elle sera inserée, auec tout le respect qui vous est dû, dans les Transactions de ce mois quoy que ie souhaiterois fort, que cette dispute fut bien terminée, et qu'on n'eut pas besoin d'entretenir les lecteurs de ces Journaux auec des particularitez, qui tesmoignent de la chaleur et de l'animosité des personnes de merite, et auec cela du soupcon de l'incertitude mesme des conoissances Mathematiques."

2.6 Concluding Remarks

Oldenburg's motivations for putting an official end to the controversy were dictated by extra-mathematical reasons, such as the desire to avoid awakening doubts in readers about the certainty of mathematics.

On the other hand, the fundamental question at the centre of the debate remained unanswered: can the circle be squared analytically? That is, is the ratio between the area of the circle and the square on its diameter expressible through a known rational or surd number? Although both Huygens and Wallis were convinced that Gregory had failed to deduce the impossibility of the definite quadrature of the circle from the impossibility of the indefinite one, their opinions diverged regarding the solvability of the circle-squaring problem.

Wallis agreed with Gregory that the ratio between the circle and the square on the diameter could not be expressed through a known number. For this reason, he qualified his objections to Gregory as "Objections against his Demonstration," instead of as objections against the latter's "doctrine." Wallis sympathized with Gregory's "doctrine," particularly with his belief in the impossibility of squaring the circle analytically, "having many years since demonstrated the same ... though [Gregory] take no notice of it, in my *Arithmetica Infinitorum*, propositio CXC with ye Scholium annexed."[107] The reference is, as we know, to the commentary to Proposition CXC of the *Arithmetica Infinitorum*.

Contrary to Gregory and Wallis, Huygens believed that the ratio between the area of the circle and the square built on its diameter (i.e., $\frac{pi}{4}$) was expressible by a rational or surd number. His idea appears to have been inspired, in the first instance, by pragmatic considerations: "The research into the quadrature of the circle—he explained in the letter of 12 November 1668—had led geometers to make so many beautiful discoveries that, in order to avoid leaving mathematicians without such a useful exercise, I consider that I must defend against Mr. Gregory the possibility of succeeding."[108] Huygens, who was familiar with classical geometry, was possibly thinking of the beautiful discoveries of the ancients, who concocted curves such as the quadratrix and the Archimedean spiral in order to find their solution to the circle-squaring problem.

Huygens' belief in the possibility of giving a geometrically exact, perhaps even numerically exact, expression for π as a rational or surd number might also have been grounded in a belief that any geometrical magnitude could be constructed in Cartesian geometry. That is, any magnitude, including the area of the circle, could be expressible as the root of a finite polynomial equation. This property of Cartesian geometry, which we

[107] Huygens (1888–1950, Vol. 6, p. 288).

[108] Huygens (1888–1950, Vol. 6, p. 272): "La recherche sur la quadrature du cercle a fait trouver tant de belles choses aux Geometres, qu'afin qu'ils ne soient pas privez d'un exercice si utile, je suis d'avis de defendre contre Monsieur Gregory la possibilité d'y reussir."

may term "completeness," was at stake at the beginning of the *VCHQ*, as Gregory wrote in the preface.

Gregory's impossibility theorem was to refute this idea of completeness by showing that the limit of the convergent sequence $\{I_n, C_n\}$ does not exist within the space of analytical quantities, although it denotes a well-defined geometrical magnitude. In light of the remarks offered in the *VCHQ*, Gregory suggested that the infinite succession of operations that engendered the sequence itself might stand as a "sixth operation." As he explained in an anticipatory remark in the preface of the *VCHQ*, this operation is infinite, and it does not coincide with any finite combination of the five arithmetic operations.

This way of conceiving incommensurable and not-analytical operations is also discussed in the *Scholium* to Proposition XI, in which it is rooted in an empiricist view about mathematical knowledge:

> It must be noticed that it is a very true philosophers' axiom, namely that all our knowledge has its beginning from the senses. Thus, among ratios, only commensurable ones are reached by the senses and are perfectly known by the human mind; an incommensurable ratio is indeed contemplated by mathematicians only in this way, to the extent that it is subduplicate, subtriplicate etc. of a commensurable ratio, or generated by the addition, subtraction etc. of such ratios. That is, a quantity which is incommensurable to a proposed quantity can be contemplated by the human mind only in this way, insofar it can be composed from the addition, subtraction, multiplication, division and root extraction of so many known quantities, and commensurable with the proposed quantity.[109]

For Gregory, therefore, the bedrock of mathematical certainty is commensurable ratios, whose knowledge is directly derived from experience. Gregory's reasons for this claim could have come from the fact that only mutually commensurable physical or geometrical magnitudes can actually be measured at all. In contrast, mutually incommensurable or non-analytical magnitudes are known only indirectly, through the mediation of known commensurable magnitudes. Here, Gregory suggests that the conceptual apparatus of the theory of proportion is necessary in order to define incommensurable ratios or irrational quantities, eschewing the reference to infinite approximation procedures. For instance, we could define an incommensurable ratio between magnitudes $a = 1$ and $b = \sqrt{2}$ as the subduplicate ratio of the commensurable ratio between magnitudes $a = 1$ and $c = 2$, since the following proportion holds: $a : b = b : c$. A similar treatment occurs for cube and higher roots. As we know, the recourse to subduplicate, subtriplcate, etc., ratios is the

[109] *VCHQ*, p. 28: "Advertendum est verissimum philosophorum axioma, nempe omnem nostram cognitionem a sensu ortum habere: inter proportiones enim, sola commensurabilis sensu attingitur & perfecte ab humana mente intelligitur; incommensurabilis enim a mathematicis solummodo adhuc contemplatur, quatenus commensurabilis cuiusdam rationis est subduplicata, subtriplicata etc. vel ex talium additione, subtractione etc. genita. hoc est, quantitas quae quantitati propositae est incommensurabilis ex eo solummodo humana mente contemplatur, quod ex aliquot quantitatum cognitarum & propositae quantitati commensurabilium additione, subductione, multiplicatione, divisione & radicum extractione componi possit."

consequence of the discovery of incommensurability, i.e., the impossibility of reducing any geometrical length to the exact ratio between two natural numbers.

The possibility of representing new magnitudes through the creation of new operations and symbolic notations beyond the ones already in use was a theme familiar to other mathematicians at the time. For instance, Wallis touched upon it in his *Arithmetica Infinitorum* (1656).[110]

According to Wallis, previous proofs had shown that one can overcome "impossible" problems in arithmetic through appropriate enrichment of the available symbolism. More precisely, such an enrichment was made possible by a "method of representing what is assumed to be done, though it may not be done in reality."[111] To take a concrete example, Wallis claimed that this idea could be used to solve the impossible equation: $3 + x = 2$ (where x is a natural numbers). Mathematicians had started by assuming that a solution existed. They then had elaborated a new notation in order to express said solution. In the case at hand, introducing negative numbers allowed for the equation to be solved. Likewise, to solve an equation that does not admit any integer solution, like $3x = 2$, mathematicians had to follow an analogous method and introduce fractions. A further analogous case was equations involving irrational roots, as: $x^2 = 2$.[112]

By reducing the circle-squaring problem to the arithmetical problem of interpolating certain number-sequences, Wallis was able to reduce the problem of computing the area of a circle with unitary radius to the same theoretical framework as was exemplified by the case of the "impossible" equations listed above. As also mentioned above, Wallis claimed, in his exchanges concerning the Gregory affair, that he had proved that the area of the circle could not be expressed as a radical or rational multiple of the radius. Although such a statement should be taken with a grain of salt.

In Gregory, we find instead a different approach to the nexus between impossibility and the extension of mathematical knowledge. Indeed, Gregory was untouched by the concern of expressing the area of the circle by new kinds of number. The impossibility of finding an analytical ratio between the circumference and the diameter (or between the circle and the square on the diameter) entailed, for him, the notion that the sole admissible measures for the area of a circle must be expressed by approximations through analytical quantities or numbers. In fact, finding better approximation formulas was indeed one of the topics of the *VCHQ* after Proposition XI.

It is also clear from the *VCHQ* that Gregory's book took Cartesian geometry as its explicit critical target. Let us recall that the *VCHQ* opened precisely with the question of whether Descartes' *Géométrie*, relying on the five arithmetical analytical operations, was endowed with sufficient generality to express any proportion at all among quantities and thus solve all geometrical problems. The impossibility result expounded in Proposition XI

[110]Panza (2005, pp. 75–78) and Stedall (2002, pp. 181–183).

[111]Wallis (2004, p. 162).

[112]Wallis (2004, pp. 162ff.).

gave a negative answer to this question. The same concern emerges once more in the final *Scholium* of the treatise, in which Gregory discussed, from a methodological viewpoint, the concept of solvability in geometry:

> By means of the analysis and of our doctrine of convergent series I can solve many of those problems which before were judged impossible to solve. However, one may perchance object that these solutions are not geometrical. I answer that if one understands by geometrical the practice requiring the sole aid of the ruler and compass, not only will these problems be impossible, but so will all the other problems which cannot be reduced to quadratic equations, as can easily be proven. If by geometrical one understands the reduction of a problem to an analytical equation, then all these problems are geometrically impossible, as it is manifest, from the results proven here, that this reduction cannot be performed. However, if by geometrical one understands the simplest of all possible methods, it may be discovered, after ripe consideration, that all the said problems are solved in the most geometrical way.[113]

This passage contains Gregory's reflections on the main foundational issues in early modern geometry: when should a problem be considered adequately solved? Which means are to be considered adequate for a problem at hand? As it appeared, at least from Descartes' *Géométrie*, answers to these questions, besides being based on mathematical considerations, were also a matter of methodological or meta-theoretical choice.

As Gregory pointed out, one could adhere to the Euclidean canon and identify geometrical exactness with solvability by ruler and compass. Or one could extend, like Descartes did, the definition of geometrical exactness to comprise anything susceptible to being solved by recourse to algebraic curves. Both such concepts of exactness were obviously too narrow for the scope of Gregory's work, which concerns problems irreducible to analytic (or algebraic) solutions. Therefore, to these criteria, Gregory added a third criterion of geometricity: geometricity is a property of the solution to a problem, which is obtained when the problem has been solved by the simplest possible method. However, perhaps this is not a criterion at all, for it leaves open how one ought to define the simplest, and therefore the more geometrical, method of solving a given problem. Certainly, impossibility results have a role to play in this endeavour, insofar as the impossibility of solving a problem by prescribed means sets a limit to simplicity "from below," so to speak, by ruling out methods that are too simple. Thus, a rule and compass alone cannot be employed to solve the trisection of an angle, nor can algebraic curves alone be employed to solve

[113] *VCHQ*, p. 58: "Multa talia problemata possem hic resolvere ope analysios & nostrae serierum convergentium doctrinae, quae antea impossibilia aestimabantur: sed dicet forte aliquis has resolutiones non esse geometricas; respondeo, si per geometricum intelligatur praxis ope solius regulae & circini peracta, hanc in his non solum esse impossibilem sed etiam in omnibus problematis quae ad aequationem quadraticam reduci non possunt, sicut facile demonstrari posset; & si per geometricum intelligatur reductio problematis ad aequationem analyticam, omnia haec problemata sunt geometrice impossibilia, cum ex hic demonstratis, manifestum sit talem reductionem fieri non posse: si vero per geometricum intelligatur methodus omnium possibilium simplicissima; invenietur fortasse post maturam considerationem omnia praedicta problemata esse geometricissime resoluta."

the quadrature of the circle. Yet, once we know which methods are too simple to be of any use, the problem of which of the possible methods is simplest remains unresolved. And, for the circle-squaring problem, we can still rely on an array of tools, ranging from transcendental curves such as the quadratrix of the ancients to various infinite series.[114] Gregory simply dismissed the question. For Gregory, simplest is that which the mathematician finds to be the simplest, depending on the circumstances and the problem the mathematician is called upon to solve. Anything goes, provided the conditions of solvability are respected. Or nearly anything, as Gregory was convinced that his method was after all the most geometrical one, and he believed that anyone would agree, even if "after ripe consideration."[115]

Returning to the quadrature of the circle, let us stress once more that no definitive and decisive argument was, in the end, deployed to debunk Gregory's impossibility claims once and for all. Even though Huygens and Wallis shared a common dissatisfaction with Gregory's arguments, one of the main points of controversy, namely, the algebraic or analytical quadrability of the circle, remained unsolved by the beginning of 1669.

In the next chapter, I look at the immediate legacy of this controversy. I claim that its unsatisfactory ending was one of the driving forces that continued to motivate, for three or four years afterward, the interest shown by Huygens for Leibniz's treatise *De Quadratura arithmetica circuli ellipseos et hyperbolae cujus corollarium est trigonometria sine tabulis*, and the impossibility arguments that can be found in the latter work.

[114]As we read in the *GPU*, p. 7: "If someone wishes to square the circle or divide an angle in a given ratio, organically, I can't see how it can be done in a manner simpler than by the vulgar quadratrix line, described in the solid and in the plane, accurately and pointwise." Gregory was probably referring to the generation of the quadratrix from the cutting of solid sections (described in Pappus' *Collectio*, Book IV, proposition 28, 29) or from a point by point construction (this can be found in Clavius, in Book VI of his second edition of the Commentary to Euclid's *Elements*, published in 1589).

[115]*VCHQ*, p. 58.

Leibniz's Arithmetical Quadrature of the Circle 3

3.1 Introduction

Leibniz set out to prove the impossibility of solving the indefinite—or, as he sometimes also chose to term it, the "general," or the "universal"—quadrature of the central conic sections in a few manuscripts written during the years 1675–1676. The clearest formulation of this impossibility theorem can be found in the last Proposition (Proposition 51) of the 1676 book-length treatise *De quadratura arithmetica circuli ellipseos et hyperbolae cujus corollarium est trigonometria sine tabulis* (hereinafter *De quadratura arithmetica*).[1]

As the full title of the treatise suggests, Leibniz's discovery is the "arithmetical quadrature" of the central conic sections, that is to say, the expression of the area of any arbitrary sector of the circle, the ellipse or the hyperbola by an infinite convergent series of rational numbers. Moreover, in speaking of a "more geometrical" quadrature, Leibniz had in mind a solution obtained by means of finite algebra, according to the canon of Cartesian geometry.

[1]Here and in the following sections, I will use the letter 'A,' followed by one Roman and one Arabic numeral, in order to refer to the edition of Leibniz's collected works published in the Academy Edition of Leibniz's miscellaneous works (Leibniz 1923). Thus, 'AVII6' will refer to the sixth volume of the seventh series of the Edition of the *Akademie der Wissenschaften*, and 'AVII6, 51' will refer to the text number 51 contained in that volume. In particular, AVII6, 51 contains a new critical edition of *De quadratura arithmetica*, with an additional passage with respect to the first edition released by E. Knobloch in 1993 (for simplicity, I will use the shorthand 'LKQ' in order to refer to Knobloch's edition. A new, recent edition of Leibniz's treatise, with a German translation, is Leibniz (2016). See also Leibniz (2004) for a French translation, and Leibniz (2015) for a Spanish translation.). Finally, I shall use the abbreviation 'LSG' for Gerhardt's historical edition of Leibniz's mathematical works published in seven volumes (1849–1863, Leibniz 1849–1863).

D. Crippa, *The Impossibility of Squaring the Circle in the 17th Century*,
Frontiers in the History of Science, https://doi.org/10.1007/978-3-030-01638-8_3

In this chapter, I shall argue that the impossibility theorem stated above comes as the result of Leibniz's critical reflection on Gregory's *VCHQ*, and on Gregory's subsequent polemics with Huygens. At the same time, Leibniz essentially clarified the meaning of the problems of the indefinite quadratures of the circle and the hyperbola, as well as their impossibility.[2] For him, the impossibility of the indefinite quadrature of the circle boiled down to the impossibility of expressing the length of an arbitrary circular arc as an analytical (in Gregory's terminology, or algebraic, in our own) function of its chord. Additionally, since Leibniz identified the hyperbolic area with the logarithmic function, the impossibility of the indefinite quadrature of the hyperbola boiled down to the impossibility of expressing the logarithm of a certain quantity as an analytical function of that quantity.

Regarding the influence exerted by Gregory upon Leibniz, the historian of mathematics Joseph Hofmann conjectured, while preparing the critical edition of Leibniz's scientific correspondence from the years 1672–1676,[3] that Leibniz's interest in the impossibility of the circle-squaring problem stemmed directly from Huygens' suggestion that Leibniz study both the *VCHQ* and the *De circuli magnitudine inventa*, perhaps in the hope that the young mathematician would be able to solve the riddle regarding the analytical quadrability of the central conic sections. The material contained in Volumes 3–6 of the Academy Edition of Leibniz's mathematical manuscripts (AVII3-AVII6), the last of which is specifically devoted to the arithmetical quadrature of the conic sections, substantially endorses Hofmann's thesis. Moreover, it shows a stronger link than previously suspected between Gregory's work and Leibniz's reflections on the quadrature of the central conic sections, and their impossibility.

In order to present Gregory's influence on Leibniz's mathematical development, I shall answer the following set of questions below: How well was Leibniz acquainted with Gregory's work (Sect. 3.3)? At which point in his research on arithmetic quadrature did Leibniz formulate and prove his impossibility results, and what was their significance (Sect. 3.8)? Which type of criticism, if any, did Leibniz raise against Gregory (Sect. 3.8.1)? Which proof-technique was adopted by Leibniz in order to prove his impossibility result on the indefinite quadrature of the circle and the other central conic sections (Sect. 3.9, and Sect. 3.9.1)? And, finally, did Leibniz manage to solve the most pressing question at stake in the controversy between Gregory and Huygens, namely, the (im)possibility of the algebraic quadrature of the whole circle (Sect. 3.11)?

[2]Another recent work exploring some of the issues dealt with in this and the previous chapter, also in connection with other contemporary impossibility theorems, in particular, Wallis and Newton, is: Lützen (2014), to which I address the reader.

[3]AIII,1, pp. LV–LVI.

3.2 The Manuscripts of *De quadratura arithmetica*

As Leibniz's *Nachlass* reveals, *De quadratura arithmetica* had a complex editorial history.[4] Leibniz composed several drafts, starting from Autumn 1673, which circulated among his friends and fellow mathematicians.[5] From the autumn of 1673 to October 1674, Leibniz also composed four Latin treatises (in chronological sequence: AVII4 42; AVII6 1, 3, 8 = AIII1 39.1) and a French treatise (AIII1, 39, sent to Huygens). From late in 1675, two French fragments are extant, one destined for La Roque (AIII1, 72) and the other supposedly for Gallois (AIII1, 73).[6] The successive manuscripts date from Spring to September 1676: AVII6, 14, 20 and 28, 51.

On the basis of the manuscripts mentioned above, we can say that Leibniz worked on the problem of the quadrature of the circle from 1673 (AVII4, 42) to September 1676, the last month of his stay in Paris. Leibniz possessed a manuscript of his treatise that was ready for publication in September 1676, a fact that he confided to his friend Soudry. The latter was to remain in Paris while Leibniz would move first to London for a brief period and then, from the end of 1676, settle down in Hannover.[7] The project of preparing the treatise for print, continued by Soudry over the next few years, came to a halt because of the loss of Soudry's own copy around 1680.[8] Leibniz, who still possessed his original Parisian manuscript (this manuscript was published in LKQ and, with some additional passages, in AVII6, 51), ultimately turned down the opportunities to publish his arithmetical quadrature in its full length. Leibniz's determining motivation was that he no longer saw the need to publish the treatise, since its content and techniques had eventually been surpassed by the method of differential and integral calculus.[9]

[4]Cf. Knobloch (1989), Probst (2006) and Probst (2008) and the introduction to the recently published AVII6.

[5]As far as is known, Leibniz first made mention of his arithmetical quadrature in a letter to Henry Oldenburg, the permanent secretary of the Royal Society, from July 15, 1674 (AIII1, 120. Cf. Knobloch 1989, p. 128).

[6]Concerning the latter, I point out that, in his *Catalogue Critique* of Leibniz's manuscripts, Rivault mentions Huygens as a recipient.

[7]AIII2, pp. 104, 105, 116, 129, 145; AII1, p. 482.

[8]See Knobloch (1989, pp. 132–133), for a detailed account.

[9]Cf: LSG, 3, p. 128: "Iam anno 1675 compositum habebam Opusculum Quadraturae Arithmeticae amicis ab illo tempore lectum, sed quod materia sub manibus crescente limare ab editione non vacavit, postquam aliae occupationes supervenere, praesertim cum nunc prolixius exponere vulgari more, quae Analysis nostra nova paucis exhibet, non satis pretium operae videret." Leibniz's motivations are perfectly clear from this passage. We could add, nevertheless, another, equally clear letter to Jacques Bernoulli, in which Leibniz stated: "I did not think that my Arithmetical Quadrature, although it was received by the French and English with great commendation, was worth being published, as I was loath to waste time over such trifles when the whole ocean was open to me" (in Child 1920, p. 20).

However, Leibniz published, starting in 1682, several articles that summarised the main results of his Parisian manuscript. I mention, in particular, the *De Vera Proportione Circuli ad Quadratum in Numeris Rationalibus Expressa*,[10] the *De geometria recondita et analysi indivisibilium atque infinitorum* (1686),[11] and the *Quadratura Arithmetica communis Sectionum Conicarum quae centrum habent* (1691), published in the *Acta Eruditorum* (See LSG, V, p. 128). Another interesting but unpublished piece is the *Compendium quadraturae arithmeticae* (LSG, V, p. 99), probably from 1690–91 (Probst 2008, p. 171).

Now that its editorial history has been discussed, let us return to the extant manuscript of *De quadratura arithmetica* (LQK or AVII6, LI). The manuscript can be divided into three main sections.[12] A first part (Propositions 1–11) contains Leibniz's attempt to provide a rigorous justification for the infinitesimal techniques on which his inquiry into quadratures is based.[13]

A second part, which extends from Proposition 12 to Proposition 25, contains a set of preparatory theorems for the quadratures of the circle and the hyperbola, which are, in fact, executed in the third part (Propositions 26–51). Indeed, from Proposition 27 to Proposition 32, Leibniz presents the arithmetical quadrature of a sector of the circle, obtained by expressing the area of the sector by means of an infinite series, and deduces what might be called, nowadays, the alternate converging series for $\frac{\pi}{4}$, namely:

$$\frac{\pi}{4} = 1 - \frac{1}{3} + \frac{1}{5} - \ldots + \frac{(-1)^n}{2n+1} + \ldots$$

Then, from Proposition 42 to Proposition 50, Leibniz dealt with the arithmetical quadrature of the hyperbola. Finally, Proposition 51, the "crowning" proposition of Leibniz's treatise, contains his result concerning the impossibility of a more geometrical quadrature of the central conic sections.

Leibniz presented his impossibility result at the end of the treatise, in Proposition 51, notably after having explicated his own method for quadratures and having exhibited a solution to the quadrature of the circle and the hyperbola. An impossibility result was not added by Leibniz to any of the drafts of *De quadratura arithmetica* until 1676. In the same year, Leibniz stated this result (without proof) in a manuscript from late spring 1676 (AVII, 6 n. 19, pp. 176–177), which was written as a preface for the *De quadratura arithmetica*, and in a tract from spring 1676 as well (AVII6, 18). Later, Leibniz inserted the impossibility result, together with a proof, into a draft of *De quadratura arithmetica*

[10]In LSG, 5, p. 120. See also: Leibniz (1989, p. 77), for a French translation.

[11]In LSG, 5, pp. 226–233; Leibniz (1989, p. 126).

[12]Knobloch (1989, p. 139).

[13]As remarked in AVII6, p. XXIV, Leibniz was employed in working out this justification himself from 1675 onwards. See, in particular, AVII6, 14, dating from early 1676, AVII6, 20, from spring 1676, and the latest version of *De Quadratura Arithmetica* AVII51, Propositions 1–11. See Knobloch (2002) and Knobloch (2008). See also the recent: Arthur (2013) and Rabouin (2015).

from late summer 1676 (AVII, 6 n. 28), and later, in the final version of his treatise, in September 1676.

3.3 Leibniz's Knowledge of Gregory's Works

The manuscript corpus presented in Volumes 3, 4, 5 and 6 of the seventh series of Leibniz's collected works shows that, during the years 1672–1676, Leibniz became incredibly conversant with Gregory's attempts to prove the impossibility of squaring the central conic sections analytically and with the subsequent controversy with Huygens. This familiarity reached the point such that Leibniz himself raised some of Huygens' previously leveled objections, albeit in a more profound and considered form.

As a copy of a letter from Oldenburg to Curtius attests, Leibniz was already informed about Gregory's *Vera circuli et hyperbolae quadratura* in 1668, before his arrival in Paris.[14] Other cursory remarks, found in a letter sent to Jean Gallois in 1672 and in a manuscript from the end of the same year, reveal that Leibniz had already been aware of the general outlines of Gregory's method of quadratures since the beginning of his Paris sojourn.[15]

Later on, it was Huygens, Leibniz's mentor during his Parisian period, who introduced Leibniz to Gregory's geometrical works in 1672–3 (cf. AVII1, 3). In fact, it is precisely recorded in Huygens' notebooks, under the date of December 30, 1673, that Leibniz borrowed a copy of the *De circuli magnitudine inventa*, a copy of Gregory's *VCHQ* and possibly the relevant letters concerning the controversy between Huygens and Gregory from Huygens (Huygens 1888–1950, Vol. 20, p. 388). On that occasion, Huygens probably instructed Leibniz about the issues at stake in that controversy and about the reasons why Gregory's arguments were not considered sufficiently cogent. Leibniz began a study of Huygens' treatise and Gregory's *VCHQ* immediately afterwards.[16]

On the other hand, a group of manuscripts now published in AVII4 (in particular numbers 31, 32) informs us about the way in which Leibniz came to know and read the *Exercitationes Geometricae*, published by Gregory in autumn 1668. The introduction of this book summarizes and discusses the considerations on the impossibility of the analytical quadrature of the circle proposed in the *VCHQ* and the criticism advanced by Huygens in his first public intervention (Huygens 1888–1950, Vol. 6, p. 229). Leibniz

[14]The letter dates from July 13, 1668. See: AII1, N. 9, pp. 17–18.

[15]See AIII, 1 p. 3, AVII3, 6, p. 65; AVII3, 20, p. 249.

[16]Early notes on Gregory's *VCHQ* can be found in AVII6, 2, 3, and in AVII5, 13. Many years later, Leibniz claimed, in a letter to Wallis (May 28 (or June 7) 1697, AIII, 7, p. 428) that he had only skimmed through Gregory's book while in Paris (this point will also be discussed below). In light of the aforementioned manuscripts, though, this recollection appears to be entirely false, since it appears that Leibniz studied at least the first part of Gregory's book, the one that deals with impossibility, with a certain degree of care.

certainly acquired Gregory's book during his first visit to London, in Spring 1673.[17] During the same visit, Leibniz bought and later perused two other mathematical works, Mercator's *Logarithmotechnia sive methodus construendi logarithmos nova, accurata & facilis* (1668) and Michelangelo Ricci's *Geometrica Exercitatio de maximis et minimis* (see Mercator 1668). Both texts were probably sold in a single volume together with Gregory's *EG*. Leibniz's volume is still preserved in Leibniz's library in Hannover.[18]

Gregory's *Geometriae Pars Universalis*, completed in 1668, was read by Leibniz only starting from the end of 1674.[19] Although Leibniz was very interested in this book, it seems that he did not work through it, but only studied a few propositions.[20] Since Leibniz's interest in this work is not directly related to impossibility results, but rather to the research on the quadrature of figures based on a method for transforming a given plane bounded region into an equivalent surface, I shall not consider the *GPU* any further in this inquiry.

The above survey thus confirms that, soon after starting his studies on the circle-squaring problem, during 1673, Leibniz had come to know about Gregory's technique of quadratures, his attempts to prove the impossibility of solving this problem analytically and the controversy that ensued on this subject between Gregory and Huygens. As a result, Leibniz's research on the circle-squaring problem proceeded in a parallel way with his interest in the controversy on the impossibility of solving the problem analytically.[21]

[17]As Gerhard recalls: "Leibniz paid two visits to London from Paris [the first] was from January 11 to the beginning of March 1673; the second was made on his way home to Germany, when he stopped in London for about a week, in October, 1676" (in Child 1920, p. 159).

[18]See Hannover, Niedersächs. Landesbibl. Ms IV 377; AVII4, 3, p. 48; AVII5, 47, p. 332.

[19]See AVII5, p. xxi.

[20]Hofmann argues that Leibniz did not possess a personal copy of this book while in Paris (Hofmann 1974, pp. 75–76). This conclusion is justified on the basis of the extant collection of books possessed by Leibniz, now in the Library of Hannover. There are, in fact, two copies of the *Geometriae Pars Universalis*; one came to Leibniz after Huygens' death (I point out that this copy contains Huygens' annotations too), which occurred in 1695, and the other originated in a similar way from Martin Knorre's library (Hofmann 1974, p. 76); (Mahnke 1925, p. 29). The copy from Knorre (catalogued as Marg. 98) also contains a reissue of the *VCHQ*, annotated by Leibniz. The annotations are obviously dated later than 1676, and may be dated to the end of the Seventeenth Century.

[21]Gregory's achievements were discussed by Leibniz on other occasions too, all of which, however, fell after 1676, and thus outside the historical bounds of this study. For instance, Leibniz was outspoken about this in a letter to Tschirnhaus, from 1684 (AIII, 4, p. 174), and in the context of a controversy on Tschirnhaus' alleged method for the integrability of algebraic curves: "Id ipsum scilicet ego objeci, verum hujus precautionis nullum in ejus edito Schediasmate reperitur vestigium, sed quia probaverat non dari quadraturam circuli portionumque ejus indefinitam, quod dudum constabat, sine haesitatione concluserat impossibilitatem quadraturae totius circuli, in quo argumentandi modo et Jac. Gregorius insignis geometra olim lapsus erat, quaemadmodum recte a viro celeberrimo Christiano Hugenio fuit observatum." Other critical references to Gregory can be found in a letter to Wallis, dated May 28 (or June 7) 1697, in which Leibniz remarked: "Ostendit mihi olim Hugenius Parisiis Jac. Gregorii perbrevem libellum in 4 in quo videbatur aliqua contineri

In order to clarify how precisely Leibniz's interest in the debate around the possibility/impossibility of the circle-squaring problem came together with his original research on the arithmetical quadrature of the central conic sections, I shall reconstruct the major results expounded by Leibniz in the set of drafts devoted to this topic written between 1673 and 1676.

3.4 The Arithmetical Quadrature of the Circle: Its Main Results

Despite Leibniz's rather dismissive remarks,[22] the secondary literature has largely focused on the first propositions of *De quadratura arithmetica*, in which Leibniz presented his method of transmutation of curves and proved its "rigorous" foundations.[23]

---

promotio serierum convergentium, sed aenigmaticos quamquam mihi inspicere tantum in transitu non legere vacarit" (AIII, 7, p. 428). Although Leibniz claims here that he only skimmed through Gregory's book, he probably read the first part, concerning Gregory's impossibility argument, as we may especially suppose in the light of the manuscripts now published, such as: AVII,6 n. 18, 28. Finally, there is a later manuscript (*Paralogsmus Jac. Gregorii cum circuli algebraice quadrari posse negat*, LH 35, XIII, 1, Bl. 118. This piece is also mentioned in Breger 1986, p. 121), which refers to a purported "paralogism" committed by Gregory. As can be read in the manuscript, the paralogism concerns Gregory's definition of convergent series, not his impossibility arguments. At any rate, its dating is still uncertain, although it was possibly written during Leibniz's stay in Hannover (it might well be related to the exchanges with Wallis from 1697). Most likely, Leibniz's annotations to Huygens' copy of *the VCHQ* (in particular, Proposition 10, 25), to be found in the Library of Hannover (Marg. Ms 98), should be dated to around the same period.

[22]"Hujus propositionis lectio omitti potest, si quis in demonstranda prop. 7. summum rigorem non desideret. Ac satius erit eam praeteriri ab initio, reque tota intellecta tum demum legi, ne ejus scrupulositas fatigatam immature mentem a reliquis, longe amoenioribus, absterreat. Hoc unum enim tantum conficit duo spatia, quorum unum in alterum desinit si in infinitum inscribendo progrediare; etiam numero inscriptionum manente finito tantum, ad differentiam assignata quavis minorem sibi appropinquare. Quod plerumque etiam illi sumere pro confesso solent, qui severas demonstrationes afferre profitentur" (AVII6, 51, p. 527). And in the *scholium* of the same proposition: "Hac propositione supersedissem lubens, cum nihil sit magis alienum ab ingenio meo quam scrupulosae quorundam minutiae in quibus plus ostentationis est quam fructus, nam et tempus quibusdam velut caeremoniis consumunt, et plus laboris quam ingenii habent, et inventorum originem caeca nocte involvunt, quae mihi plerumque ipsis inventis videtur praestantior" (AVII6, 51, p. 533).

[23]Leibniz's proof has been studied as an early example of Riemann's integration. For example, in Knobloch (2002), we read that: "Leibniz demonstrated the integrability of a huge class of functions by means of Riemannian sums which depend on intermediate values of the partial integration intervals." Or, again, in Levey (2008), it is claimed that: "The demonstration of Prop. 6 articulates a general technique for finding the quadrature of any continuous curve that contains no point of inflection and no point with a vertical tangent ... the technique itself is also of interest, for Leibniz's use of 'elementary' and 'complementary' rectangles very precisely anticipates Riemannian integration." In my opinion, such an emphasis on the Riemannian character of Leibniz's method could have the drawback of cutting the bonds between Leibniz and other contemporary mathematicians who were also engaged in the search for rigorous foundations for quadrature

On the other hand, less attention has been given to another fundamental result of *De quadratura arithmetica* (or, at least, one among the most "*ameniores*," to use Leibniz's terminology) expounded later on, in Proposition 43. There, Leibniz proved a "rule for the general quadrature of the conic sections having a determined centre."[24] Such a rule for the general (or indefinite) quadrature of the central conic sections can be summarized in the form of the following theorem:

Theorem 3.1 *Let $CAFE$ be a sector of a circle, an ellipse, or an hyperbola with centre C and semi-conjugate diameter AC. Let AT be a segment cut, upon the tangent to the curve at the point A, by another tangent to the same curve at the point E (Fig. 3.1). Let the conjugate semiaxis be $BC = 1$, and $AT = t$, $t \leq 1$, then the area of the sector $CAFE$ will be equal to a rectangle whose sides have length equal, respectively, to CB and to the convergent series: $t \pm \frac{t^3}{3} + \frac{t^5}{5} \pm \frac{t^7}{7} + \ldots$. The symbol $\pm$ should be interpreted as a sum, in the case of the hyperbola, or a difference, in the case of the circle and the ellipse.*

From this general result, Leibniz derived the following corollary:

Corollary 3.1 *The Circle is to the circumscribed Square, or the arc of a quadrant to its diameter, as $\frac{1}{1} - \frac{1}{3} + \frac{1}{5} - \frac{1}{7} + \frac{1}{9} - \frac{1}{11} \ldots$ to the unity.*[25]

This corollary establishes the value of the ratio $\frac{\pi}{4}$, namely, the "true proportion of a circle to the circumscribed square, expressed in rational numbers," as the title of a famous paper by Leibniz recites.[26]

I assume that a good entry point for understanding how Leibniz arrived at these results is not *De quadratura arithmetica* itself, but rather the earlier drafts of the treatise from 1674 (namely: AIII1, 39, AVII6, 4) and 1675 (AIII1, 72, 73). The main reason for this is that Leibniz privileged, in his ultimate version of the treatise, a stricter deductive ordering and a synthetic presentation of his results, and reformulated them using the somewhat more cumbersome apparatus of the theory of proportions. By providing a justification for

methods such as Gregory himself (one can consult, for example, the long discussion in cf. Whiteside 1961, pp. 331ff.). Here, and Leibniz was no exception to this, the appeal to "rigour" should be referred to the model of *reductio* proof found in the classical method of exhaustion. That Leibniz had this in mind is confirmed by the *scholium* to Proposition 7 and by the reference to Archimedes as his source of inspiration (cf. Rabouin 2015, pp. 359–360).

[24]AVII6, 51, Proposition 43 (p. 618); cf. also the later articles *Quadratura arithmetica communis*, in Leibniz (2011, p. 69); Leibniz (1989, p. 173).

[25]*De quadratura arithmetica*, XXXII: "Circulus est ad Quadratum circumscriptum, sive arcus Quadrantis ad Diametrum ut $\frac{1}{1} - \frac{1}{3} + \frac{1}{5} - \frac{1}{7} + \frac{1}{9} - \frac{1}{11}$ etc. ad unitatem ", AVII6, 51 p. 600. Cf. also: AIII1, 39, p. 165; AVII6, 4, p. 74; 7, p. 89.

[26]This is the "De vera proportione circuli ad quadratum circumscriptum in numeris rationalibus expressa," which appeared in the *Acta Eruditorum* in 1682. See Leibniz (2011, p. 7).

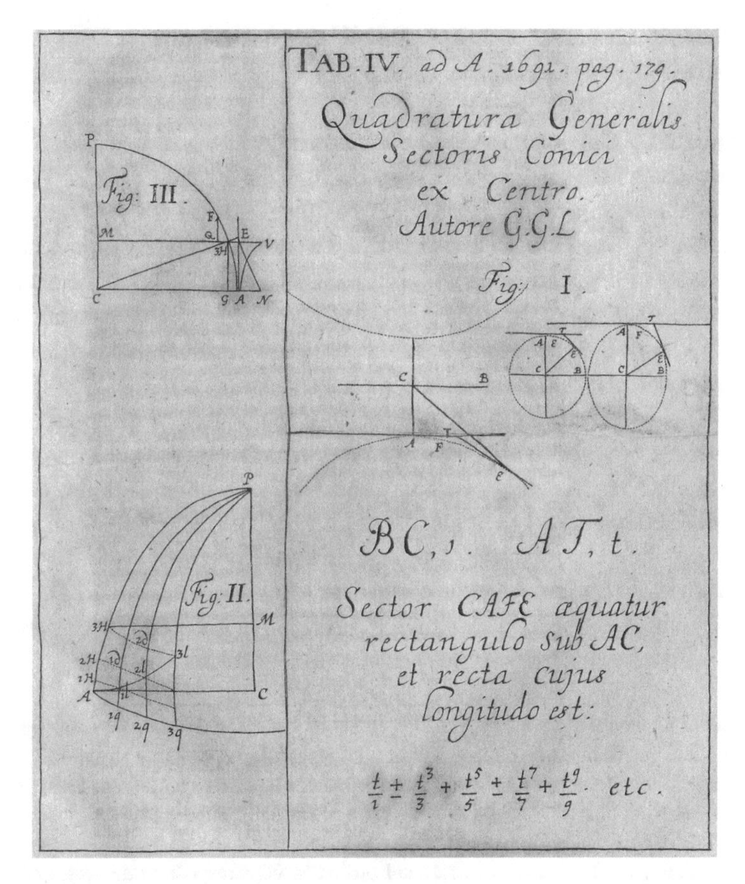

Fig. 3.1 A sector $CAFE$ of a hyperbola, a circle and a hyperbola, respectively. From Leibniz's article: *Quadratura arithmetica communis sectionum conicarum, quae centrum habent*, published in *Acta Eruditorum*, 1691, p. 179. Permission granted by Gottfried Wilhelm Leibniz Bibliothek–Niedersächsische Landesbibliothek, Hannover (shelf mark: Aa-A 35, 1692, Taf. IV)

his discoveries in terms of the structure and proof method of classical Greek geometry, Leibniz probably intended to endow his treatise with a solid and classical architecture, which might make it more palatable to his recipients at the *Académie des Sciences*, for instance. However, the synthetic style has the defect of concealing the original process of discovery and the possible influences exerted by other mathematicians such as Gregory and, as we shall see, Mercator or Wallis, on Leibniz's techniques of quadrature. Moreover, the 1674 and 1675 versions were the first samples of the treatise to circulate among Leibniz's correspondents, eliciting some initial critical responses. Some of these reactions were instrumental in Leibniz's decision to add a closing impossibility result to the final version of the treatise.

The technique used by Leibniz to solve the quadrature of the central conic sections differed from Gregory's approach in the *VCHQ* in one substantial respect: whereas the latter avoided infinitesimals and relied on a variant of Archimedes' polygonal approximation (as we have seen in the previous chapter), Leibniz used infinitesimal techniques, at least in the drafts of *De quadratura arithmetica*.

This important distinction apart, Leibniz conceived of Gregory's and his own study of quadratures as both belonging to a separate realm of mathematics called "Archimedean geometry," after the name of its ancient and most illustrious master. This type of geometry, Leibniz contended, was cultivated during the Seventeenth Century, when old methods such as the study of the centre of gravity or the method of indivisibles were rediscovered and improved upon by several contemporary mathematicians.[27] The mathematicians who undertook this work included Galileo and his disciples Cavalieri and Torricelli, but also James Gregory, Mercator, Brouncker, and Wallis.

Since 1673, Leibniz had been convinced that this realm of geometry lay beyond Descartes' geometry, which was of little or no use when it came to problems of quadrature and rectifications.[28] In particular, Leibniz conceived of his own research as improving and generalizing the work of the other "Archimedean" mathematicians on problems of quadrature.

In a long text that was originally planned as a preface to his 1676 *De quadratura arithmetica*, entitled *Dissertatio exoterica*,[29] Leibniz commented in these terms on the legacy of the British mathematicians, especially Brouncker and Mercator:

> The Viscount Brouncker, president of the Royal Society, was the first to my knowledge to give an arithmetical quadrature of a definite sector of a hyperbola, as we know, by an infinite series of rational numbers. In very recent times, the German Nicolaus Mercator, excellent geometer, gave a very elegant universal arithmetical quadrature of the sectors of the hyperbola. But he was not ready to express the dimension of the circle and its parts by an infinite series of rational numbers. In fact, the hyperbola has rational ordinates with respect to the asymptote, from which came the discovery of Mercator, but the the circle will yield irrational ordinates in any way it is treated. However, since I have discovered a very general theorem, by whose means any figure can be easily transformed into another different but equivalent one, I concluded that

[27]Let us recall that Leibniz was convinced that Archimedes used indivisibles to discover theorems about the quadratures and cubatures of curvilinear figures or solids of revolution (AVII6, 49_1, p. 498: "Archimedem ego semper in tantum miratus sum, in quantum licet mortalem; usque adeo insignia ejus inventa, et profunda, et superioribus dissimilia, et in omnem posteritatem valida fuere. Indivisibilia certe, aut si mavis infinite parva, Geometriae sublimioris clavem, adhibuit primus, tecte licet, et ita, ut admiratio inventis, et rigor demonstrationibus constaret)."

[28]AIII, 1, p. 139: "Monsieur Descartes a travaillé après Viète, à reduire les questions de Geometrie, aux resolutions de Equations, dont le calcul est entierement Arithmetique. Mais ny luy ny Viète n'ont touché qu'aux Questions Rectilignes, c'est à dire dans les quelle son ne cherche ny suppose que la grandeur de quelques lignes droites, ou figures rectilignes, à quoy se resuident en effet tous les problemes plans, solides, sursolides, etc. Mais il faut avouer que les Problemes curvilignes sont d'une toute autre Nature, et qu'on peut dire, qu'elles sont ny plans, ny solides, mais de l'infinitesiême degrez, si on les vouloit resoudre par le moyen des Equations."

[29]The full title of this tract is: *Dissertatio exoterica de usu geometriae, et statu praesenti, ac novissimis ejus incrementis* (AVII6, 49_1).

> one must see whether the circle may not, at some point, be converted into a rational figure. But this occurred indeed, in a beautiful way, as I shall expound at more length in this volume.[30]

The above passage sums up very well the rationale and motivations behind Leibniz's arithmetical quadrature of the central conic sections. Leibniz sought to express the indefinite quadrature of the circle through an infinite series of rational numbers, thus in a way similar to the arithmetical quadratures of the hyperbola obtained by Brouncker, and later by Mercator. Another source for Leibniz was Walli's interpretation of Mercator's treatise, as I shall discuss below.

However, the analytical techniques that worked so well for the quadrature of the hyperbola of equation $y = \frac{1}{1+x}$ were not immediately applicable to the case of the circle, since its equation, in the form: $y = f(x) = \sqrt{1 - x^2}$, involved expressions under a square root (see, for instance, AVII6, 49_1, p. 510).

Leibniz managed to circumvent this problem by means of a geometric transformation of the original figure, a sector of a circle, into another figure whose area stood in rational proportion to the given sector. By means of this method of transformation, referred to in the aforementioned *dissertatio exoterica* as "*theorema generalissimum*," Leibniz was eventually able to apply the method of Mercator to the newly obtained figure, and to deduce thereby the arithmetical quadrature of the circular sector. The same transformation applied to the other central conic sections yields their respective quadratures.

3.5 A Digression: Early Modern Techniques for the Quadrature of the Hyperbola

From our perspectives, the problems of the quadratures of the circle and the hyperbola are connected, because both their solutions depend on transcendental functions or transcendental numbers. Starting in the second half of the Seventeenth Century, early-modern geometers also considered both problems together. Their similarity was initially based on rather practical grounds: both quadratures appeared to be difficult, or perhaps impossible, to solve by means of elementary geometry or algebra. More specifically, as we have seen

[30] AVII6, 49_1, p. 510: "Vicecomes Brounckerus, Societatis Regiae Anglicanae Praeses primus quod sciam certae portionis Hyperbolae Quadraturam dedit Arithmeticam, per infinitam scilicet Seriem Numerorum rationalium. Novissime Nicolaus Mercator Germanus, Geometra egregius omnium in universum Hyperbolae partium Arithmeticam dedit Quadraturam perelegantem. Circuli autem et partium ejus Magnitudinem infinita numerorum rationalium serie exprimere non ita promtum erat; nam Hyperbola ordinatas in Asymptoton habet rationales, unde Mercatoris inventum fluxit, Circulus autem utcunque tractetur irrationales praebet ordinatas. Ego vero cum Theorema quoddam generalissimum reperissem, cujus ope quaelibet figura in aliam plane diversam sed dimensione aequipollentem converti potest, experiundum statui, an non Circulus aliquando in rationalem transmutari posset figuram, quod vero tandem pulchre successit, quemadmodum in sequente opusculo fusius exponam."

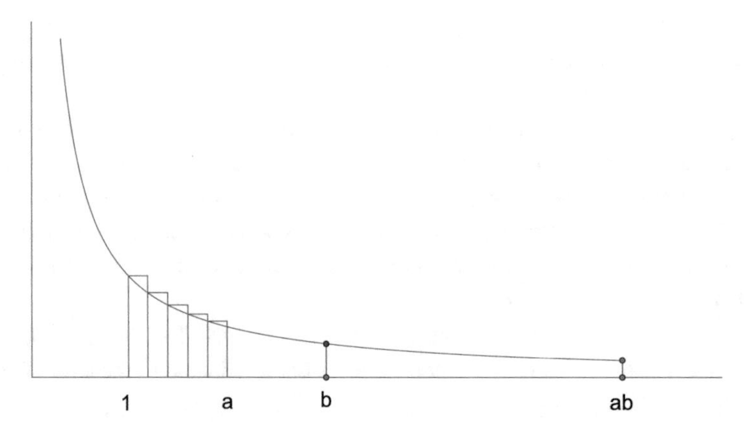

Fig. 3.2 In the hyperbola $y = \frac{1}{x}$, we can show that the area of the sector delimited by the segment $a - 1$ is equal to the sector delimited by the segment $ab - b$. To prove this, let us approximate the former area by n rectangles of equal length, as in the figure. It results that the corresponding approximation by n rectangles to the sector delimited by the segment $ab - b$ has exactly the same area. If we imagine that the number of equally-spaced rectangles tends to infinity, relying on an intuitive idea of limit, we can conclude that the sectors are equal one to the other. A more rigorous proof, according to the Greeks' ideal of rigour, could be given using an Archimedean-like indirect method based on a double *reductio* proof

in the previous chapter, Gregory related the unsolvability of the circle- and hyperbola-squaring problems by known, algebraic methods to the necessity of extending the bounds of geometry with new operations. Thus, the area of the circle and that of the hyperbola became, from Gregory's perspective, the first examples of non-analytical quantities.

Erstwhile, the hyperbola-squaring problem had acquired a central place in the mathematics of the second half of the Seventeenth Century, mainly in connection with the study of logarithms. The discovery of this relation, which dated before the onset of calculus, can be traced back to A. de Sarasa's work *Solutio problematis a R. P. Marino Mersenno Minimo propositi* from 1649, which, in its turn, built on Grégoire of Saint Vincent's *Opus Geometricum Quadraturae Circuli et Sectionum Coni* from 1647.[31]

The proof that the hyperbola provides a continuous representation of the logarithmic relation is elementary. Let us consider, on the hyperbola $y = \frac{1}{x}$, on which a unity segment 1 has been constructed (Fig. 3.2), the sectors delimited by the segment $a - 1$ and $b - 1$. We shall have that the sector delimited by the segment $ab - 1$ is equal to the sum of the previous sectors. If we call $A(a)$ the area of the first sector (delimited by $a - 1$), $A(b)$ the area of the second sector (corresponding to $b - 1$), and $A(ab)$ the area of the third sector,

[31] cf. Burn (2001), Cajori (1913), esp. p. 11.

we will have that the following relation is satisfied: $A(a) + A(b) = A(ab)$. But this is, by definition, a logarithmic relation.

To prove the crucial result that $A(a) + A(b) = A(ab)$, we can advance the following consideration. The sector delimited by the segment $ab - 1$ (that is, $A(ab)$) is the sum of the sectors delimited by segments $b - 1$ (namely, $A(b)$) and $ab - b$. We will then prove that the latter sector is equal to the sector delimited by $a - 1$ (namely, $A(a)$). The proof can be carried out without using calculus, resorting, for instance, to a double-reductio proof-method in the style of the ancients (see Fig. 3.2).

Let us then imagine constructing, under the hyperbola $y = \frac{1}{x}$, two equal, adjacent sectors. If the first sector is constructed upon a segment a, then the sum of the two will be $2A(a) = A(a^2)$. In other words, the sector equal to the sum of the given sectors will cut a segment with length a^2 on the horizontal asymptote. If we add a third equal sector, we shall have a new segment a^3 on the horizontal, and so on. In other words, if we can construct an arithmetic progression of hyperbolic areas, we shall have a geometric progression of base-segments, and vice versa. Thus, our hyperbola-model includes that which was taken to be the standard basis of logarithm construction during the first half of the Seventeenth Century, namely, the matching of an arithmetical with a geometrical progression.[32]

By the second half of the Seventeenth Century, the practical significance of the hyperbola-squaring problem had become evident: if one had a method of computing the areas of the hyperbolic sectors, one would also have a means of computing the logarithms of the corresponding base-segments. After James Gregory published the *VCHQ* at the end of 1667, Brouncker, Mercator and Wallis all made efforts to solve the quadrature of an equilateral hyperbola by means of the newly introduced technique of infinite series, and, in this way, to provide a satisfactory solution to the problem of computing the logarithms of arbitrary positive quantities.

Leibniz read both Brouncker's and Mercator's works on the hyperbola in 1673.[33] Aside from these readings, Leibniz was also acquainted with the account of Mercator's *Logarithmotechnia* given by Wallis in two letters from July and August 1668, both of which appeared in the *Philosophical Transactions*.[34] As I shall argue below, Wallis's review had an influence on Leibniz's arithmetical quadrature of the circle. Finally, in addition to these texts, Leibniz certainly knew the geometrical interpretation of Mercator's method for the quadrature of the hyperbola given by James Gregory in his *Exercitationes geometricae* from 1668.[35]

[32]Cf. Burn (2001, p. 4) and Cajori (1913, p. 7).

[33]See VII3, 6 and 8, for Mercator, and for Brouncker, see VII4, 36, from Summer 1673.

[34]cf. Wallis (1668).

[35]See *N. Mercatoris quadratura hyperbolae geometrice demonstrata*, in Gregory (1668b).

3.5.1 Brouncker's Quadrature of the Hyperbola

Brouncker's result in the quadrature of an equilateral hyperbola had been known since the early 1650s,[36] but it was published in a brief article only on April 13, 1668, in the *Philosophical Transactions of the Royal Society*. This publication came out the same year as Mercator's book *Logarithmotechnia: sive methodus construendi logarithmos nova, accurata, et facilis* (*Logarithmotechnia: or a new, accurate, and easy method of constructing logarithms*, 1668), which contained an appendix with an arithmetical solution to the quadrature of the hyperbola as well.[37]

Both Brouncker's and Mercator's works, together with Wallis's review and Mercator's response, appeared in the same year. This proliferation of publications on the same subject well illustrates the general interest that the hyperbola-squaring problem had aroused among mathematicians.

Brouncker's quadrature relies on a finite method, which consists in approximating to the area under a sector of equilateral hyperbola $ABCEA$ (Fig. 3.3) by sequences of rectangles and triangles in a way I now detail.[38]

Let us consider the hyperbola $y = \frac{1}{x}$, passing through points E, δ, and C (Fig. 3.3), and let the segment $AB = 1$ on the asymptote be repeatedly bisected.

On the basis of this iterative construction, Brouncker succeeded in expressing the areas of the hyperbolic sector $ABCEA$, the area of its complement $EDC\delta E$, and that of the hyperbolic segment $E\delta C$ through an infinite series of rational numbers.

Brouncker considered the hyperbolic surface $ABCEA$ to be the infinite sum-sequence: $S_1 + S_2 + S_3 + \ldots$, where each term is a rectangle obtained by the repeated bisection of the base segment AB, as shown in Fig. 3.3. Analogously, he considered the area of the complement $EDC\delta$ to be the infinite sum-sequence of rectangles: $A_1 + A_2 + A_3 + \ldots$. Finally, he represented the segment $E\delta C$ as an infinite sum-sequence of triangles (see Fig. 3.4).

By positing $OA = 1$, the areas of the small rectangles can be easily computed numerically. The area of the first rectangle is $S_1 = \frac{1}{2}$, while the area of S_2 is $\frac{1}{3\times4}$. By elementary computations, we can calculate the areas of further rectangles: $S_3 = \frac{1}{5\times6}$, $S_4 = \frac{1}{7\times8}$, etc., which offer better approximations to the sector itself.

By virtue of the recursive geometric procedure that generates the sequence of rectangles, the area of the sector $ABCEA$ shall be expressed by an infinite series of rational

[36] Stedall (2008, p. 84).

[37] Mercator's work was first published in August 1667, but was then republished the following year with an appendix containing the quadrature of the hyperbola.

[38] The most complete exposition of Brouncker's quadrature method can be found in Zeuthen (1903). Cf. also Hofmann (1974, pp. 96–97) and Whiteside (1961, pp. 222–223).

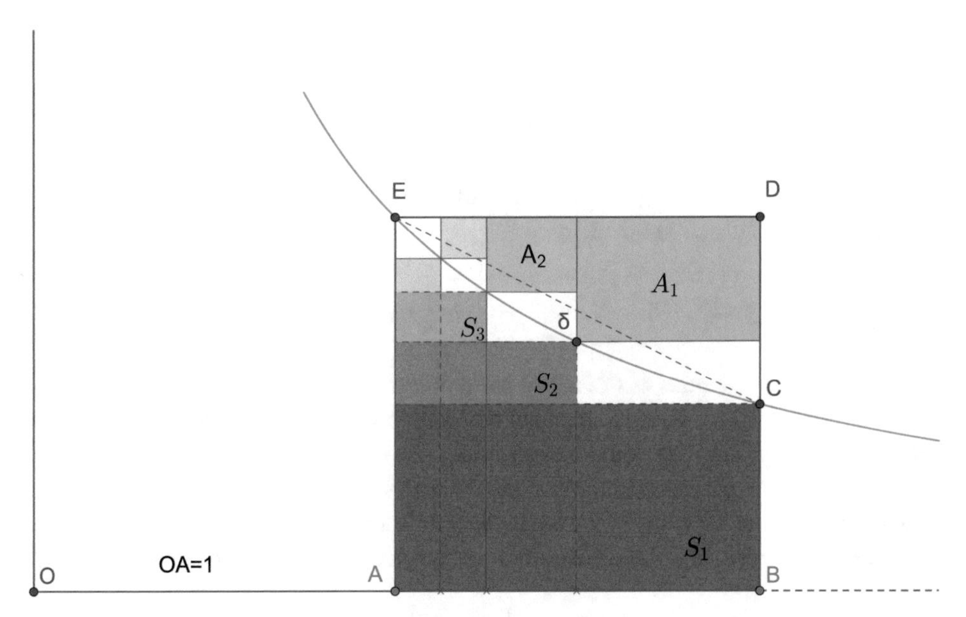

Fig. 3.3 Brouncker's quadrature of a sector under the hyperbola $y = \frac{1}{x}$. Through successive bisections of the segment AB, we can obtain a sequence of inscribed rectangles forming a polygonal figure (below the curve) inscribed in the hyperbolic sector $ABCEA$, and a sequence of rectangles forming a polygonal figure (above the curve) inscribed in the sector $EDC\delta E$

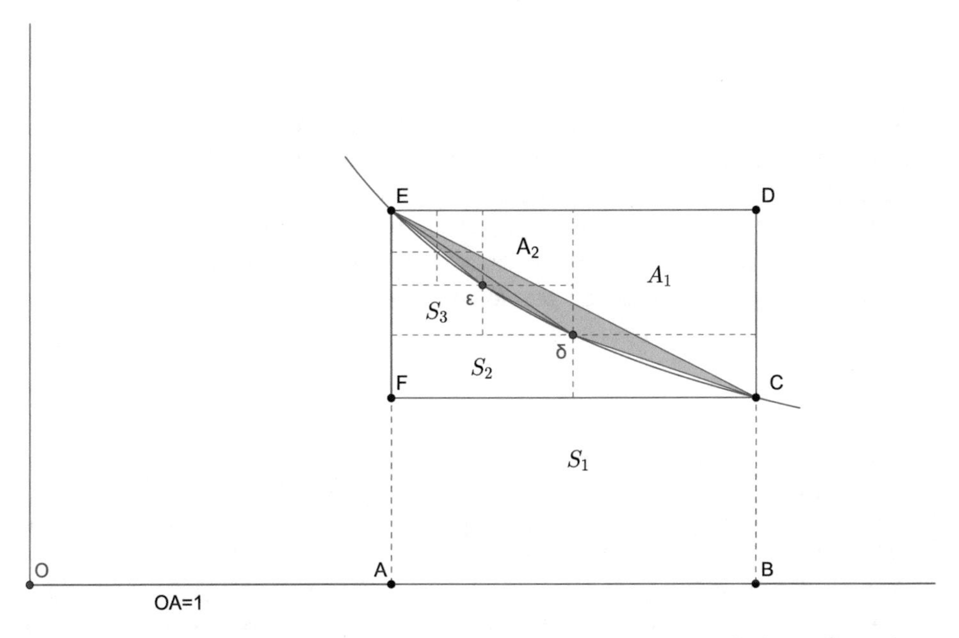

Fig. 3.4 Brouncker's quadrature of the hyperbolic segment $E\delta C$. The area of this curvilinear figure is filled with triangles, such as in the figure

numbers, namely:

$$ABCEA = \frac{1}{2} + \frac{1}{3 \times 4} + \frac{1}{5 \times 6} + \frac{1}{7 \times 8} + \frac{1}{9 \times 10} + \frac{1}{11 \times 12} + \dots \tag{3.1}$$

This series, which also gives the value of $log(2)$, can be further simplified using elementary arithmetic, so as to obtain

$$ABCEA = (1 - \frac{1}{2}) + (\frac{1}{3} - \frac{1}{4}) + (\frac{1}{5} - \frac{1}{6}) + \dots \tag{3.2}$$

Brouncker did not state this result in his publication, but he was probably aware of it, as can be derived by a simple manipulation of the original series. As we shall see in Sect. 3.10, Leibniz was to obtain the same expression for the area of the hyperbolic sector by applying a different method.

Similarly, Brouncker computed the area under the sector $EDC\delta A$ as the infinite sum-sequence of rectangles $A_1, A_2, \dots$ (Fig. 3.3), which gives the following numerical series:

$$EDC\delta = \frac{1}{2 \times 3} + \frac{1}{4 \times 5} + \frac{1}{6 \times 7} + \frac{1}{8 \times 9} + \dots \tag{3.3}$$

Finally, the decomposition of the segment $E\delta C$ into triangles yields the following sequence of rational numbers[39]:

$$E\delta C = \frac{1}{2 \times 3 \times 4} + \frac{1}{4 \times 5 \times 6} + \frac{1}{6 \times 7 \times 8} + \dots \tag{3.4}$$

Brouncker's arithmetical quadrature of the hyperbola was, in Leibniz's opinion, the first example of a "definite" quadrature of a central conic sector achieved by an infinite numerical progression of rational numbers.

In the example discussed in the article from the *Philosophical Transactions* and reported above, Brouncker's quadrature obtains only the particular sector determined by the segment $AB = 1$. However, this procedure can be generalised, as Brouncker himself

[39] Brouncker did not give any hint about the procedure he used to derive this series, although it does not seem obvious at first sight. Following Zeuthen's suggestion (Zeuthen 1903, pp. 309–310), I reconstruct Brouncker's reasoning in the following way. Relying on Fig. 3.4, we have that $E\delta C = \frac{1}{2}(EDC\delta - E\delta CF)$. On this basis, we can express the area of the segment as the difference between two infinite series, namely: $E\delta C = \frac{1}{2}((\frac{1}{2\times3} + \frac{1}{4\times5} + \frac{1}{6\times7} + \frac{1}{8\times9} \dots) - (\frac{1}{3\times4} + \frac{1}{5\times6} + \frac{1}{7\times8} + \frac{1}{9\times10} + \frac{1}{11\times12} \dots))$. Subtracting term by term, we can obtain an infinite series expressing the area of $E\delta C$. Brouncker interpreted the new series geometrically, identifying each term with the area of a triangle inscribed in the segment $E\delta C$.

indicated in the closing line of his article, to other "hyperbolic spaces, whatever be the *rational* proportion of AE to BC."[40]

3.5.2 Mercator's and Wallis's Quadrature of the Hyperbola

The quadrature of the hyperbola stood as one of the most interesting results of Mercator's *Logarithmotechnia* (1668) as well. This work is mainly dedicated to the study of logarithms, and particularly to the presentation of: "a new, accurate and easy method" for their construction, as declared in the subtitle of the book (... *"methodus construendi logarithmos nova, accurata, et facilis"*).

Although the *Logarithmotechnia* is a difficult and, at times, obscure text, it was praised by contemporary mathematicians on a number of occasions.[41] For instance, Gregory believed that Mercator's method of computing logarithms was much better than his own (AIII, 88, p. 439), and both Wallis and Leibniz had words of praise for Mercator's approach to quadrature problems.

As for Leibniz's acquaintance with the *Logarithmotechnia*, in light of several quotations found in his manuscripts (cf. AVII6, 51, *Scholium* to Proposition 29; VII6. 41, p. 438, 49, p. 510), we can conjecture that he took up the study of that work with the aid of Wallis's reviews. Such an influence is visible, for instance, from the summary of Mercator's achievements that Leibniz gave in the manuscript AVII6, 41, titled *De operis argumento et auxiliis*:

> Mercator greatly improved the matter [namely, the problem of quadratures] in an utterly novel and very elegant way: he considered that a fractional number could be expressed by an infinite series of integers, such that $\frac{1}{1+x}$ is equal to the quantity: $1-\frac{x}{1+x}$, and $\frac{x}{1+x}$ is equal to $x-\frac{x^2}{1+x}$, and $\frac{x^2}{1+x}$ to $x^2 - \frac{x^3}{1+x}$ and so on, and then once all the terms had been collected $\frac{1}{1+x}$ is equal to the series: $1-x+x^2-x^3\ldots$ Indeed, by virtue of the arithmetic of the infinite, the sum of all 1 is the last abscissa ("*novissima*") x; and the sum of all x is the last abscissa ("*novissima*") $\frac{x^2}{2}$. Indeed let there be a curve, like the hyperbola, whose abscissa is x, and ordinate $\frac{1}{1+x}$ or,

[40] Brouncker (1668, p. 648).

[41] On the difficulty of the text, I point to Gregory's opinion, as it is recounted by Collins in his *Historiola*: "he [Gregory] betook himself to the study of the said Logarithmotechnia, but complained he could not understand the Authors meaning being so involved about Phrases concerning Ratios" (AIII1, 88, p. 440). The project undertaken by Gregory to explain Mercator's quadrature of the hyperbola in terms of the rigorous proof-structure of exhaustion method can be viewed as the author's attempt to dispel the obscurities of the *Logarithmotechnia* (cf. Gregory (1668b) and Malet 1996, p. 60). A similar aim might have been behind Wallis's reviews.

therefore, $1 - x + x^2 - x^3$ etc. the sum of all the previous ordinates, or the area of the figure, will be $\frac{x}{1} - \frac{x^2}{2} + \frac{x^3}{3} - \frac{x^4}{4}$ etc, as it is known from the quadratures of the parabolas.[42]

In the above passage, Leibniz pointed out what he considered to be two salient aspects of Mercator's solution of the hyperbola-squaring problem that made the latter a novel and outstanding achievement.

The first aspect concerns the algorithm for expanding a rational polynomial such as $\frac{1}{1+x}$ (which could also be a fraction, if x takes a numerical value) into an infinite series, also known as the method of long divisions. According to Leibniz's summary of the procedure (the original can be found in *Logarithmotechnia*, Proposition XV), the algorithm initializes with the explicit division of 1 by the binomial $1 + x$, and continues by taking, at each step, n as the fractional remainder $R_n(x) = \frac{a_n(x)}{1+x}$ and dividing the expression $a_n(x)$ by the binomial $(1 + x)$.

In the case at hand, we shall thus have the following infinite division[43]:

$$\begin{aligned} \frac{1}{1+x} &= 1 - \frac{x}{1+x}, \\ \frac{x}{1+x} &= x - \frac{x^2}{1+x}, \\ \frac{x^2}{1+x} &= x^2 - \frac{x^3}{1+x}, \\ \frac{x^3}{1+x} &= \dots \end{aligned} \tag{3.5}$$

The second aspect, which is emphasized in Wallis's review of the *Logarithmotechnia*, consists in using, for the quadrature of the hyperbola, a well-known result pertaining to the quadrature of higher parabolas (namely, the family of curves whose equation is: $y = ax^k$,

[42] AVII6, 41, p. 438: "Mercator diversa plane ac pereleganti ratione rem longius produxit: consideravit enim numerum fractum exprimi posse serie integrorum infinita, (a) ut $\frac{1}{1+x}$ esse aequale quantitati: $1 - \frac{x}{1+x}$ et $\frac{x}{1+x}$ aequari huic $x - \frac{x^2}{1+x}$ et $\frac{x^2}{1+x}$ huic $x^2 - \frac{x^3}{1+x}$ et ita porro, ac proinde omnibus collectis aeqvari $\frac{1}{1+x}$ seriei $1 - x + x^2 - x^3$ etc. (aa) Jam per arithmeticam infinitorum summa omnium 1 est x novissima; et summa omnium x est $\frac{x^2}{2}$ novissima (bb). Sit jam curva cuius abscissa sit x, ordinata $\frac{1}{1+x}$, qualis est Hyperbola erg. | vel $1 - x + x^2 - x^3$ etc. erit summa omnium ordinatarum praecedentium seu area figurae, $\frac{x}{1} - \frac{x^2}{2} + \frac{x^3}{3} - \frac{x^4}{4}$ etc ut notum est ex quadraturis parabolarum."

[43] The same process is explained, in *De quadratura arithmetica*, by taking an arbitrary fraction $\frac{a}{b+c}$ (AVII6, 51, p. 596).

with $k \neq -1$). This result can be summarised thusly:

$$\int_0^{\xi} ax^k \, dx, = \frac{1}{k+1} a\xi^{k+1}.$$

As we know, the formula gives the area of the higher parabola $y = ax^k$ between the abscissas 0 and ξ (the *ultima abscissa*, in the terminology of early modern mathematicians).

We might read an analogous result through the lines of Wallis's *Arithmetica infinitorum*, published a few years earlier than Mercator's work. However, as we have briefly indicated in the previous chapter, such an analogy should be taken with a grain of salt. In Wallis's *Arithmetica infinitorum*, there is never a question of solving a quadrature problem by calculating the area of a plane figure using numbers or algebraic symbols at all. On the contrary, all of the results on quadratures are expressed via the classical formalism of the theory of proportions.

The key result that Wallis was to use in dealing with Mercator's quadrature of the hyperbola is expounded in the *Arithmetica infinitorum*, and can be summarised as follows.[44] Let $\sum_{i=0}^{h} i^k$ be a partial numerical sum (with the exponent k initially ranging among positive natural number). Wallis was able to determine, after performing a few calculations on examples involving several positive integer values of k, a result that we can express as follows[45]:

$$\frac{\sum_{i=0}^{h} i^k}{h^k(h+1)} = \frac{1}{k+1} + r(h).$$

In the above expression, which is a modern rendering of Wallis's original conclusions, $r(h)$ is a remainder that grows smaller as the number of terms increases. Generalizing the above result to an infinite number of terms, Wallis extrapolated the following theorem, which he stated for k being an integer, a fraction or a surd number, with the exclusion of $k = -1$[46]:

$$\frac{\sum_{i=0}^{h} i^k}{h^k(h+1)} = \frac{1}{k+1}. \tag{3.6}$$

From the above, one can derive:

$$\frac{\sum_{i=1}^{h} i^k}{h^{k+1}} = \frac{1}{k+1}. \tag{3.7}$$

---

[44] See Wallis (2004) (*Introduction*) and Panza (2005, Chapter 1), for a technical survey of Wallis's treatise.

[45] See Wallis (2004, pp. 13–15); p. 39.

[46] Cf. Wallis (1668, p. 758), and Proposition 64 of the *Arithmetica Infinitorum*.

Let us stress that Wallis held Eq. (3.7), which I have rewritten in a modern notation following Panza (2005, pp. 55–56), to be valid if h is an actually infinite number.[47] Wallis's reasoning appears to be defective for its lack of logical rigour in extrapolating to the infinite from finite cases, as was already duly noted in the Seventeenth Century.[48] However, despite its logical flaws, it led to important breakthroughs, such as in the case of the quadrature of the hyperbola that I examine now.

Mercator's quadrature of the hyperbola combined both aspects listed above: (1) the expansion of a rational polynomial into an infinite series and (2) the quadrature of high-order parabolas.[49] We shall present it here according to the interpretation found in Wallis's reviews, which appeared in the *Philosophical Transactions* in July and August 1668, since these reviews arguably influenced Leibniz the most.[50]

In discussing the *Logarithmotechnia*, and, in particular, those propositions dealing with the quadrature of the hyperbola "in a subtle and ingenious way" ("*elegant admodum atque ingeniosa*"), Wallis referred to the same figure found in Mercator's treatise (Fig. 3.5).

Thus, let a rectangular hyperbola FBM be given with centre A and asymptotes AE and AN. By positing $AI = BI = 1$, and $IH = a$, the curve will have the equation: $FH = y = \frac{1}{1+a}$.[51] Let an arbitrary sector $BIru$ be taken, and its basis Ir divided into infinitely many equal intervals ("*aequales partes innumeras*"). By this operation, we understand that the segment has been divided into an actual infinity of equal parts, namely: "α, 2α, 3α ... *usque ad* A," as Wallis said, where A denotes, in this case, the length of Ir. From now on, in order to avoid confusion, I shall not follow Wallis's usage, but will rather indicate the length of Ir by means of the less ambiguous letter X.

Resorting to Cavalieri's terminology, Wallis thus stated that the ordinates corresponding to each subdivision of the segment Ir "fill in" (the verb employed is "*complere*") the hyperbolic sector $BIru$.[52] Thus, the area of $BIru$ can be thought of as the "aggregate" (*aggregatum*) of infinitely small segments, each being equal to: $\frac{1}{1+\alpha}, \frac{1}{1+2\alpha}, \frac{1}{1+3\alpha}, \dots \frac{1}{1+h\alpha}$ (where h denotes an actual infinite number) from I to r. If one would not wish to consider an area filled with segments, then Wallis would have had no qualms in considering thin rectangular strips instead, with heights equal to $\frac{1}{1+\alpha}$, $\frac{1}{1+2\alpha}$, $\frac{1}{1+3\alpha}$, $\dots$ $\frac{1}{1+h\alpha}$, each base

[47] As observed by Malet and Panza, in Malet and Panza (2015, p. 313), Wallis treated the infinite as a positive integer number, which was to obey the same arithmetical operations as finite numbers.

[48] For a recent discussion of Wallis's philosophy of mathematics and its criticism, see Alexander (2014, pp. 258–278); and Jesseph (1999, pp. 173–188).

[49] Mercator expounded his method for squaring the hyperbola in Propositions XIV–XIX of the *Logarithmotechnia*. Cf. Rosso (2014) for a detailed mathematical analysis.

[50] My reconstruction is mainly based on the useful (Edwards 1994).

[51] Wallis (1668, p. 753). The reference is to *Logarithmotechnia*, Proposition XV.

[52] Wallis (1668, p. 753).

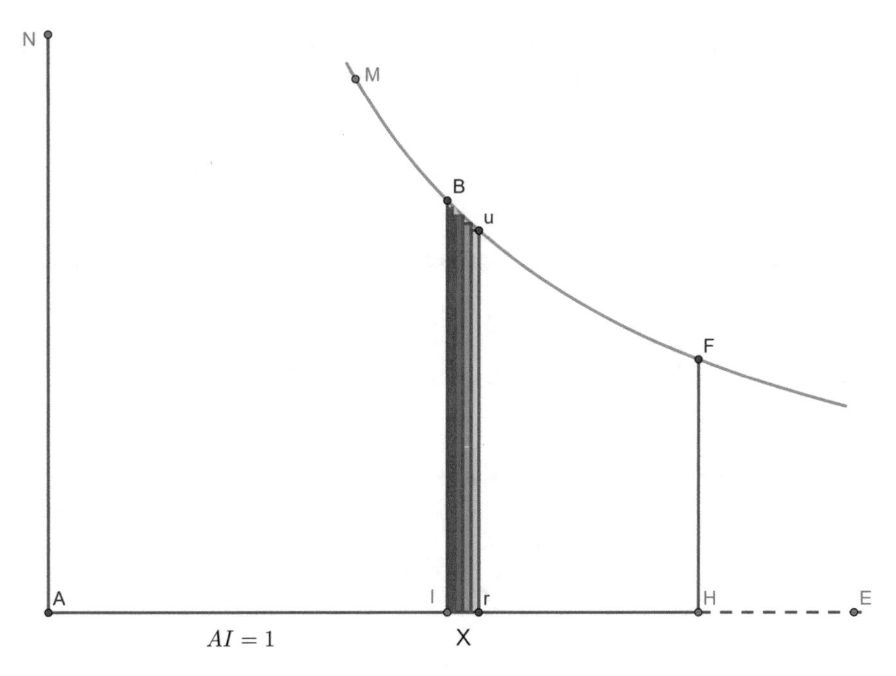

Fig. 3.5 Mercator's quadrature of the hyperbolic sector $BIru$. The sector is "filled up" with small rectangular strips with basis-length equal to α

being equal to the small segment α, namely[53]:

$$BIru = \sum_{i=1}^{i=h} \alpha(\frac{1}{1+i\alpha}).$$

[53]Cf. Edwards (1994, pp. 162–163). Let us note that Wallis had adopted, as a mathematician, a rather tolerant attitude towards indivisibles in his previous works (esp. *De Sectionibus Conicis Nova Methodo Expositis, Tractatus* (1655) and in his *Arithmetica infinitorum* (1656)) by admitting either the Cavalierian interpretation that consisted in considering the surface of a plane region as an aggregate of parallel lines, or the interpretation that consisted in considering a surface to be filled with infinitely many parallelograms with small base. For instance, he wrote in his treatise on conic sections: "I suppose from the beginning (after Bonaventura Cavalieri's *Geometriam indivisibilium*) that any [portion of a] plane consists, as it were, of infinitely many parallel [straight] lines, or rather (as I would prefer) of infinitely many parallelograms equally high, the altitude of each of which is $\frac{1}{\infty}$ of the total altitude, that is, an infinitely small aliquot part (for "∞" denotes an infinite number), so that the altitude of all [such parts] taken together is equal to the altitude of the figure" (Pasini 1993, pp. 45–46). Wallis justified his scarce interest in taking a side with respect to the question about the real nature of the indivisibles on the grounds that both interpretations evoked in the passage above are operationally equivalent. Cf. Malet (1996, pp. 23–24) and Malet and Panza (2015).

The second step in Wallis's method consists in expanding each fraction: $\frac{1}{1+i\alpha}$ $(i = 1, 2, 3, \ldots h)$, by means of Mercator's algorithm seen above, so as to obtain

$$BIru = \alpha[\sum_{k=0}^{\infty}(-1)^k\alpha^k] + \alpha[\sum_{k=0}^{\infty}(-1)^k(2\alpha)^k] + \ldots + \alpha[\sum_{k=0}^{\infty}(-1)^k(h\alpha)^k].$$

Collecting the terms containing equal powers of α, we obtain the following infinite sum for the area of $BIru$:

$$\alpha\sum_{i=1}^{i=h}(i\alpha)^0 - \alpha\sum_{i=1}^{i=h}(i\alpha)^1 + \alpha\sum_{i=1}^{i=h}(i\alpha)^2 + \ldots (-1)^k\alpha\sum_{i=1}^{i=h}(i\alpha)^k + \ldots \tag{3.8}$$

Without further specification, Wallis wrote the results of the sums $\sum_{i=1}^{i=h}(i\alpha)^k$ as

$$\sum_{i=1}^{i=h}(i\alpha)^0 = X,$$

$$\sum_{i=1}^{i=h}(i\alpha)^1 = \frac{X^2}{2},$$

$$\sum_{i=1}^{i=h}(i\alpha)^2 = \frac{X^3}{3},$$

$$\sum_{i=1}^{i=h}(i\alpha)^3 = \frac{X^4}{4}$$

$$\vdots$$

and stated that the area of the hyperbolic sector $BIru$ is (given $Ir = X$)[54]:

$$BIru = X - \frac{X^2}{2} + \frac{X^3}{3} - \frac{X^4}{4} \ldots \tag{3.9}$$

Wallis's underlying reasoning might be reconstructed in the following way.[55]

[54] Wallis (1668, p. 754).

[55] I follow here the proposal advanced in Edwards (1994, pp. 162ff.).

The formula (3.8) can be rewritten as

$$BIru = \alpha \sum_{i=1}^{i=h} i^0 - \alpha^2 \sum_{i=1}^{i=h} i^1 + \alpha^3 \sum_{i=1}^{i=h} i^2 + \ldots (-1)^k \alpha^{k+1} \sum_{i=1}^{i=h} i^k \ldots$$

from which there follows, by positing $\alpha = \frac{X}{h}$ (since $Ir = X$),

$$BIru = \frac{X}{h}h - \frac{X^2}{h^2}(\sum_{i=1}^{i=h} i^1) + \frac{X^3}{h^3}(\sum_{i=1}^{i=h} i^2) + \ldots (-1)^k \frac{X^{k+1}}{h^{k+1}}(\sum_{i=1}^{i=h} i^k) \ldots$$

Then, using Proposition LXIV of the *Arithmetica infinitorum* (cf. Eqs. (3.6) and (3.7)) one can conclude that:

$$BIru = X - \frac{X^2}{2} + \frac{X^3}{3} - \ldots (-1)^k \frac{X^{k+1}}{k+1} \ldots$$

which is Eq. (3.9) above.

Wallis claimed that the validity of the above result depended on the condition that: $Ir < 1$. He argued in the following terms: "if one queries the quadrature of a sector $BIHF$ (whose side IH is understood to be longer than AI) this procedure will not be successful: because the remedy, as we have said, will not be sufficient to cure the disease. Since, in fact, we must posit: $A > 1$ [$X > 1$, according to the letters used above]; it is evident that its successive powers will become greater, hence they should not be neglected."[56]

One way to circumvent this restriction is presented in the closing lines of the article, where Mercator's procedure is generalized to sectors with arbitrary base-length. As an example, let us compute the area of the sector $HFur$. Let us set $AH = 1$ and $Hr = X$. As usual, let Hr be divided into equal parts α. Since $AH = 1$, the successive subdivisions are the terms of the following decreasing sequence: $1-\alpha, 1-2\alpha, 1-3\alpha, \ldots 1-X = (Ar)$.

Having set: $r(AH, HF) = r(Ar, rU) = r(AI, IB) = \ldots = b^2$, we can infer: $HF = b^2$, and the successive segments, perpendicular to the division points of the base Hr, shall be: $\frac{b^2}{1-\alpha}, \frac{b^2}{1-2\alpha} \ldots$.

Since each term is of the form: $b^2(\frac{1}{1-n\alpha})$, it can then be expanded using Mercator's formula, while paying attention both to the changed sign and to the constant b^2.

Eventually, applying the same method discussed above, collecting the terms with equal power and integrating term by term, the area of the sector $HFur$ is equal to the following

[56] Wallis (1668, p. 754): "...Siquis totius spatii $BIHF$ (cujus latus IH longius intelligatur quam AI) quadraturam postulet: propter medelam, quam modo diximus, malo minus sufficientem. Cum jam ponendum sit $A > 1$; manifestum est, posteriores ipsius potetastes, altius in Integrorum sedes penetraturas, adeoque minime negligendas."

infinite series:

$$HFur = b^2(X + \frac{1}{2}X^2 + \frac{1}{3}X^3 + \frac{1}{4}X^4 \ldots). \quad (3.10)$$

Before turning to Leibniz's arithmetical quadrature, let us point out two reasons why Mercator's and Wallis's method for the squaring of the hyperbola represented a conceptual turning point in the history of quadrature problems. A first novelty can be appreciated when considering, for example, the way in which Wallis had dealt with quadrature problems in the *Arithmetica infinitorum*. In that work, Wallis had still adhered to the classical idea according to which to square a curvilinear figure S meant determining a proportion between S, some other figure (in general, a circumscribed parallelogram or a rectangle) and two integer numbers, so as to derive a procedure for constructing, by ruler and compass, a rectangle equal to S.[57]

On the other hand, in the 1668 review of the *Logarithmotechnia*, the squaring of a hyperbolic sector had ceased to have the above, classical meaning. Squaring the hyperbolic sector $BIru$ meant, by then, to express its area by means of an infinite sum-series: $X - \frac{X^2}{2} + \frac{X^3}{3} - \frac{X^4}{4} + \ldots$.[58]

Wallis did not delve at all into the nature of this series. However, the following interpretation can be ventured. Since X denotes the "last abscissa," it certainly denotes a segment. Moreover, Wallis might have employed the segment X to measure the surface of a rectangle of base-length X and height 1. This would be a natural interpretation on the basis of the usual formula for calculating the area of a rectangle. Similarly, "$\frac{X^2}{2}$" could have denoted a segment measuring the area of a triangle the base and height of which are both X.

Wallis could then have assumed, on the basis of the quadratures of higher parabolas obtained in the *Arithmetica infinitorum*, that the segment $\frac{X^3}{3}$ measured the surface of a parabolic sector cut off by the segment Ir under the parabola of the equation $y = x^2$, drawn, or imagined to be drawn, with I as a vertex. On the same basis, he might have taken segment $\frac{X^4}{4}$ to measure the surface of a trapezoid cut off by Ir under a cubic parabola of the equation $y = x^3$ (traced with its origin in I). A similar interpretation can be extended to the other terms appearing in the sum-series: $X - \frac{X^2}{2} + \frac{X^3}{3} - \frac{X^4}{4} + \ldots$, each being a segment that measures the area of a corresponding sector delimited by a higher parabola $y = x^n$. In conclusion, the whole sum-series would denote an infinite sum of segments,

[57] Panza (2005), especially pp. 58–60.

[58] A similar remark is advanced by Jean Dhombres, with regard to Wallis's review: "The word 'quadrature' lost a large part of its meaning, in the sense that no square with an equal area as the area of a hyperbolic segment is seen. Finding a series seems to be the new name for quadrature" (Dhombres 2014, p. 27).

and thereby a segment in its turn (supposing, as Wallis himself did, that the sum-series converges) that measures the area of the whole sector.[59]

A second, important change concerns the nature of infinite series and their use in the problem of quadratures. Unlike in Gregory's and Brouncker's quadrature methods, in which infinite series are extrapolated from polygonal recursive constructions according to a traditional Archimedean model, Mercator and Wallis constructed infinite series starting from a given finite polynomial and iterating the usual arithmetic operations.[60]

Wallis was convinced that the use of the so-called method of long divisions provided a quadrature so "complete and fast" ("*absoluta est tamque expedita*") that he did not know whether one could ever expect a better one to be developed.[61] This opinion reminds one of Leibniz's statement, in the closing proposition of *De quadratura arithmetica*, about the impossibility of ever finding better and more geometrical quadratures of the conic sections than his own.

This similarity of phrasing probably did not occur by chance. In fact, Leibniz knew Wallis's article well, and even quoted the passage under examination in one of the drafts of *De quadratura arithmetica*.[62] Moreover, as we shall see in the next section, an examination of the sources will reveal the deep influence that both Mercator's quadrature and Wallis's review had upon the development of Leibniz's own method of quadrature.

3.6 Leibniz's Arithmetical Quadrature: The Geometrical Reduction

As I have mentioned above, in his manuscript notes from the years 1673–1676, Leibniz repeatedly stressed the continuity between the arithmetical quadrature of the hyperbola obtained by Brouncker, Mercator and Wallis, and his own quadrature of the conic sections. At the same time, he also stressed a major difficulty that ran counter to the possibility of routinely applying Mercator's analytical procedure to the quadrature of a circular sector:

> There is no one who cannot perceive how easily this expedient has been applied to the hyperbola (…) But in fact no one, I think, dared even to hope that the circle could be treated

[59]Wallis's change in the way of dealing with quadratures reveals similarities with the approach that Newton had developed in his notebooks from 1664–1665 (cf. Panza 2005, Chapter 3). However, Wallis's knowledge of Newton's early mathematical research at the time is not documented.

[60]I substantially agree here with Ferraro and Panza (2003, p. 19) that infinite series were conceived of as "quasi-polynomials" or "infinite extensions of polynomials," entities that could be manipulated by applying the same operations applicable to finite polynomials. Cf. also Collins's words in the *Historiola*, when he described the series of Mercator as: "an Equation consisting of an infinite number of tearmes Potestates of severall degrees more properly called an infinite Series" (AIII1, 88, p. 439).

[61]Wallis (1668, p. 756).

[62]AVII6, 1, p. 30: "insignis Geometra, Joh. Wallisius … pronuntiaverit … eam esse tam absolutam tamque expeditam Hyperbolae quadraturam, ut nescire se profiteatur an meliorem sperari debeat."

> in this way. It seems to me that the following difficulty hinders us the most from getting a very easy solution to the circle squaring problem: the fact that it is impossible to free from the radical sign the ordinates directed from the curve to any axis, provided their measure is expressed in relation to the abscissas. On the contrary, it is possible in the Parabola, and in the hyperbola, and in all the simple paraboloids and hyperboloids.[63]

Remarks like this occur frequently in Leibniz's notes between 1673 and 1676.[64] These remarks all point to what Leibniz considered a difficult problem: Mercator's method of long division could be fruitfully applied to square a hyperbola, a parabola or higher parabolas (that is to say, those curves with equation: $y = ax^m$, with m integer), and, more generally, to square regions bounded by curves having rational ordinates corresponding to rational abscissas.[65] Despite these successes, the circle represented one of the most blatant (and most interesting) cases to which Mercator's algorithm of the long divisions cannot be applied, because an irrational expression emerges as soon as the ordinates of a circle are expressed in terms of their respective abscissas.

By the end of 1673, Leibniz had found out a way to circumvent the problem.[66] His solution, as can also be very plainly seen from several drafts of *De quadratura arithmetica* until the end of 1675, can be summed up in two main steps. On the basis of Leibniz's own terminology, I shall call the first step "geometrical reduction," and the second one "analytical solution."[67]

The geometrical reduction consists, as the name indicates, in reducing the problem of squaring a circular segment delimited by an arc and the subtending chord to the quadrature

[63]AVII6, 1, p. 31: "Caeterum nemo est qui non videat facilem hujus artificii ad Hyperbolam fuisse applicationem (...) At vero Circulum ipsum ita tractari posse, nemo opinor vel sperare ausus est. Ego cum ad commodam Circuli dimensionem illud maxime obstare viderem, quod ordinatae ex curva ejus ad axem alium quemcunque demissae valore per relationem ad abscissas expresso nunquam absolvi possent ab irrationalitate, cum contra in parabola et Hyperbola omnibusque paraboloeidibus et Hyperboloeidibus simplicibus possint ..."

[64]See AVII4, 36, p. 596, AVII6, 41, p. 438, AVII6, 49_1, p. 510, AVII6, 51, pp. 567, 641; AIII1, 39_2, p. 168.

[65]In *De quadratura arithmetica*, Leibniz called these curves "analytically simple." See, for instance, AVII6, 1, p. 31, n. 41, p. 438, and, in particular, AVII6, 51, p. 561: "*Curvam Analyticam simplicem* voco, in qua relatio inter ordinatas et portiones ex axe aliquo abscissas, aequatione duorum tantum terminorum explicari potest; sive in qua ordinatae earumve potentiae, sunt in multiplicata, aut submultiplicata directa, aut reciproca ratione; abscissarum, potentiarumve ab ipsis, vel contra" (the emphasis is in the original).

[66]AVII4, 27, p. 493. See also Mahnke (1925, p. 41).

[67]Such a methodological partition does not mirror the organisation of the final draft of *De quadratura arithmetica*, but corresponds to a thematic partition more evidently present in the first known couple of manuscripts of this treatise, dating from Autumn 1673 (AVII4, 42_1, 42_2), and in the excerpts sent to Huygens, in October 1674, and to La Roque and Gallois one year later (these are, respectively, the manuscripts: AIII1, 39, 72, 73).

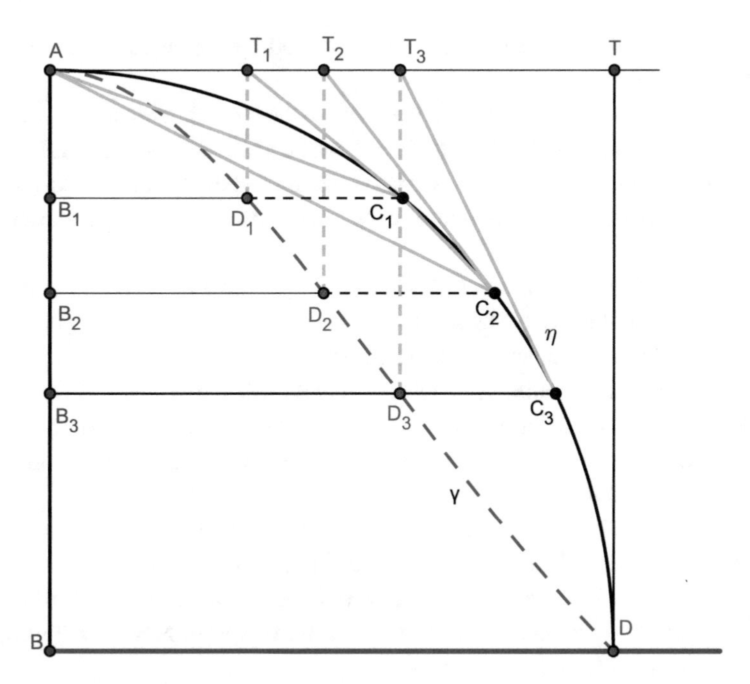

Fig. 3.6 Construction of the *figura segmentorum*

of an associated *figura segmentorum*, namely, a figure delimited by an algebraic curve, by the radius itself and by the axis of the abscissas.[68]

According to Leibniz's explanations,[69] the companion curve to a given circular sector ABC_3A (Fig. 3.6) can be constructed as follows. With vertex in point A, let a right angle BAT be constructed. Let a sequence of points C_i be arbitrarily chosen on the circumference and, from each of these points, let lines parallel to AT be drawn, meeting AB at the corresponding points B_i. Thus, from each point C_i, let the tangents C_iT_i be produced so as to meet AT at corresponding points T_i (Leibniz calls each segment AT_i "resecta," namely, an abscissa cut by the tangent passing through the point C_i). From each T_i, let the perpendicular to AT be dropped so as to meet the corresponding segments C_iB_i at a point D_i. The points D_i form the locus of the companion curve γ.

Subsequently, Leibniz proved (cf. Theorem 3.2 below) that the region bounded by the region AC_2A, including by the corresponding arc AC_i and the chord joining its

[68]In Leibniz's corpus of manuscripts concerning quadrature problems, an explicit definition of the *figura segmentorum* can found in the following passages (among others): AVII6, 4, p. 53, AVII6, 8, p. 94, AVII6, 20, p. 202, AVII6, 51, p. 539. On the other hand, the analogous expression *Figura resectarum* occurs in AVII6 51, p. 535.

[69]cf. AIII1, 39, p. 155; AVII6, 1, p. 5.

extremities, is half the corresponding region bounded by the curve γ and by axes AB_i and B_iD_i (Fig. 3.6).

As I shall discuss below, this reduction presents a clear computational advantage, since, through it, Mercator's method of long divisions can be applied to the seemingly intractable quadrature of a circular sector. This novel application of Mercator's technique stood at the core of Leibniz's arithmetical quadrature of the circle, and can be easily generalised to the other central conic sections as well.

During his investigations between 1673 and 1676, Leibniz used two different methods in order to prove that the area of the figure $ABiDi$ is twice the area of AC_2A. Thus, in the first manuscripts of the arithmetical quadrature up to the version sent to Huygens in 1674, he usually employed the technique of momenta and centres of gravity. He had originally found and mastered this technique through Pascal's geometrical work (in particular, the *Traité du sinus du quart du cercle*) and through the work of Honoré Fabri, in particular, the *Synopsis geometrica*.[70]

Mostly after 1674, he relied instead on a geometrical technique for the transformation of area, which he referred to as the method of "transmutation" (*transmutatio*). Leibniz thought highly of this technique, as he considered it a crucial advance toward the general solution of quadrature problems. For instance, in his answer to Newton's *Epistola Prior*, he explained this method in broad lines, as follows:

> My method is just a corollary of a general theory of transformations, by whose aid any given figure, describable by any equation, is transformed into another equivalent (*aequipollentem*) analytical one (...) the basis of the [method of] transformation is this: that a given figure, with innumerable lines drawn in any way (provided they are drawn according to some rule or law), may be resolved into parts, and that the parts—or others equal to them—when reassembled in another position or another form compose another figure, equivalent (*aequipollentem*) to the former or of the same area even if the shape is quite different.[71]

As is made clear in the above passage, the method of transmutation builds upon the so-called "distributive" approach, which Cavalieri employed in his technique of quadratures.[72] Cavalieri's method consists in comparing the indivisibles of two figures singularly (that is to say, each indivisible belonging to a figure is compared to a

[70]Hofmann (1974, p. 51) and Horvath (1983), in which Leibniz's early calculation with momenta is discussed in detail.

[71]AIII1, 89, pp. 569–570: "Mea Methodus corollarium est tantum doctrinae generalis de transformationibus cujus ope figura proposita quaelibet quacumque aequatione explicabili transmutatur in aliam analyticam aequipollentem (...) transformationis fundamentum hoc est, ut figura proposita rectis innumeris utcunque (modo secundum aliquam regulam sive legem) ductis resolvatur in partes, quae partes aut aliae ipsis aequales, alio situ aliave forma reconjunctae aliam componant figuram priori aequipollentem seu ejusdem areae, etsi alia longe figuram constantem." Cf. also Edwards (1994, p. 245).

[72]For a more detailed account, see Giusti (1980), Andersen (1984) and Jullien (2015).

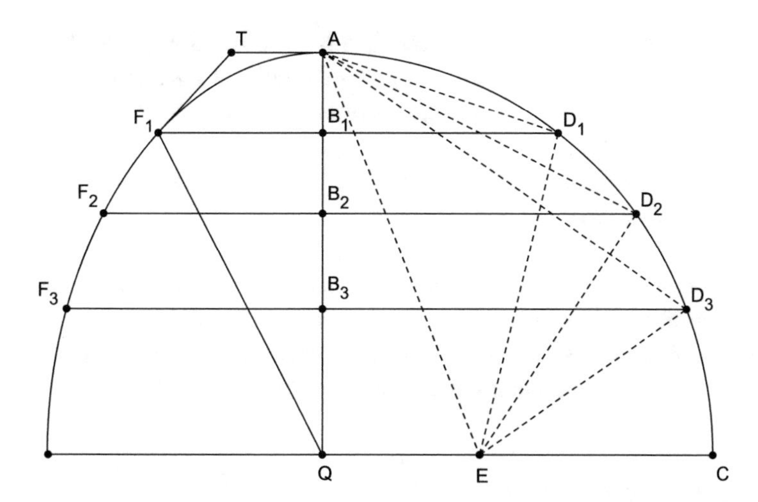

Fig. 3.7 Leibniz provided the following example in order to illustrate his transmutation method: "if the parallels BD be traced, [the figure $AQCDA$] can be decomposed [*resolvi*] into trapezoids B_1D_2, B_2D_3 etc., but if we trace convergent straight lines ED it can be decomposed into triangles ED_1D_2, ED_2D_3 etc. Now, if $AF_1F_2F_3$ is another curve, whose trapezoids B_1F_2, B_2F_3, are respectively equal to the triangles ED_1D_2, ED_2D_3, in the same order, the whole figure $AED_3D_2D_1A$ shall be equal to the whole figure $AF_1F_2F_3B_3A$" (AIII1, 89, p. 571)

corresponding indivisible of the other figure), and then inferring, from the constant ratio between each corresponding pair of indivisibles, the ratio between the two figures.[73]

However, Leibniz's method innovated upon Cavalieri's, insofar as he considered curvilinear figures not only decomposed into rectangles, but also into converging triangles, as it was the case with Roberval's and Pascal's techniques of quadrature.[74]

The fact that variants of the technique depicted in Fig. 3.7 were widespread among mathematicians from the second half of the Seventeenth Century has suggested a possible influence exerted by Barrow or by Roberval on Leibniz's own method.[75] However, the thesis of Leibniz's independent discovery has recently gained stronger confirmation in the available manuscripts, as argued in VII7, 4 (Introduction), in Probst (2015), and in the recent Probst (2016).

[73]In Cavalieri's own terms: "Arbitrary plane figures placed between the same parallels, in which, when an arbitrary line parallel to the above is drawn, the ratio between the intersections of the two figures with the line is the same as that of the intersections with any other line (the homologous terms being always in the same figure) have the same ratio (*proportionem*) between them as the said intersections" (Quoted in Jullien 2015, p. 49).

[74]See Rabouin (2015, p. 352).

[75]Besides the above-mentioned authors, we find examples of the transmutation method in Van Heuraet and Gregory as well, both of whom were certainly known to Leibniz (Mahnke 1925, p. 10).

The core of Leibniz's method of transmutation can be best appreciated by considering the way in which it is concretely used to prove the rational relation between a circular segment and the corresponding *figura segmentorum* (Fig. 3.6), namely:

Theorem 3.2 *The area of a circular segment AC_iA (Fig. 3.6) is half of the corresponding trapezoid AD_iB_i (with $i = 1, 2, 3\ldots$).*

Starting from the draft of *De quadratura arithmetica* sent to La Roque and continuing in the final draft from 1676, Leibniz articulated a proof of the above statement around two key ideas. The first idea is presented as a lemma (AIII1, 72, p. 340; AIII1, 73, p. 360; AVII6, LI, p. 523), which propounds a transmutation of a given triangle into an equivalent rectangle:

Lemma 3.1 *Through the vertices of a given triangle BEF (Fig. 3.8), let three parallels BD, GE and HF be drawn, and let the side EF be extended until it meets the parallel BD in C. The rectangle with sides GH and HK is the double of the triangle BEF.*

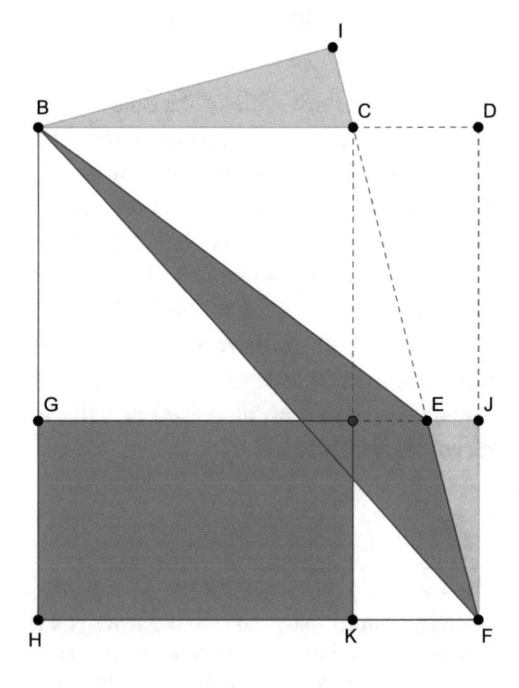

Fig. 3.8 The figure reproduces the original in the draft of the treatise sent to La Roque (AIII1, 72, p. 341)

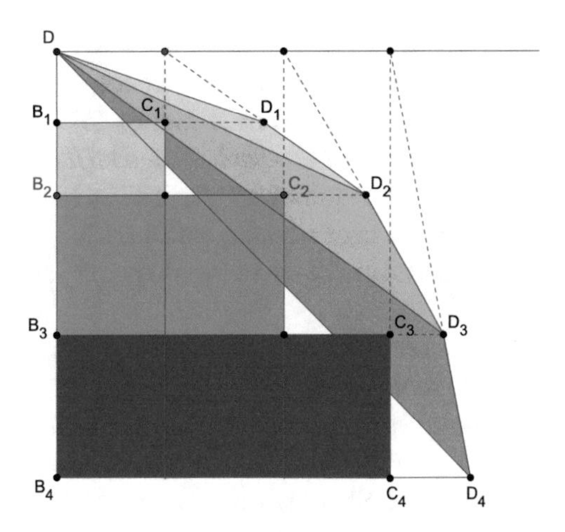

Fig. 3.9 First application of the Lemma: the polygonal area $DD_1D_2D_3D_4$ formed by concurrent triangles and the rectangular step-space $B_1B_4C_4C_3C_2C_1$ are equivalent

This lemma is, according to Leibniz himself, "easy to prove by means of ordinary geometry" ("aisé de le demonstrer par la géométrie ordinaire", AIII1, 72 p. 341). In fact, its proof depends only on the similitude between the right-angled triangles BIC and EJF (see Fig. 3.8). As a consequence, the following proportion holds: $BC(= HK) : BI = EF : JF(= GH)$. Thus, the rectangle with sides HK and GH is equal to the rectangle with sides BI and EF. The second rectangle, however, equals twice the triangle BEF (this follows immediately from the formula for the area of a triangle, since EF is the base of BEF and BI its height), thus the lemma is proven.[76]

This same lemma can be applied to measure the area of a polygonal surface, as in Fig. 3.9. In fact, it is sufficient to divide the polygonal surface into triangles having a common vertex, and then transform each triangle into its corresponding rectangle. Eventually, the polygonal area will be equal to the step-space marked in the figure.

Leibniz did not discuss the particular problem of computing the area of a polygonal figure, but directly applied the above lemma to proving the equivalence between a circular segment and its corresponding *figura resectarum*.

The application of the lemma to curvilinear surfaces was made possible thanks to the second key idea underlying Leibniz's quadrature method, namely, that of considering any arc as a polygonal line with infinitely many sides. The definition of a curve as an infinite-sided polygon became the standard definition of curves in Leibniz's *Nova Methodus* and, later on, in the first textbooks on calculus.[77]

[76]The proof can be found, for instance, in AVII6, LI, p. 523. See also Rabouin (2015, p. 351).

[77]Thus, L'Hôpital considered it as a postulate, in his *Analyse des infiniment petits* (1696): "that a curve line may be considered as the assemblage of an infinite number of infinitely small right lines: or (which is the same thing) as a polygon of an infinite number of sides, each of an infinitely small length, which determine the curvature of the line by the angles they make with each other." Quoted in Mancosu (1999, p. 152).

In Leibniz's view, such a conception of curves as polygons with infinitesimal sides went hand in hand with a new conception of tangent. This novelty clearly appears in an early group of notes from August 1673, entitled *De functionibus*. In particular, in the tract called *Methodus nova investigandi tangentes linearum curvarum ex datis applicatis*, Leibniz expressed the view that curves could be understood as: "consisting of infinitely many straight lines or sides, which are like portions of the tangents joining two proximal points applied on the curve (or two points separated by a distance infinitely small between them."[78]

Given these premises, Leibniz's proof of Theorem 3.2 goes as follows. Let us consider, with reference to Fig. 3.6, the segment C_1C_2 of infinitesimal length on the circumference, and construct the triangle AC_1C_2. Let us then trace the ordinates B_1C_1 and B_2C_2. Since points C_1 and C_2 are chosen sufficiently close to one another, namely, at an "infinitesimal" distance, as Leibniz would say (cf. AIII, 1, 72, p. 341), the cord C_1C_2 will become indistinguishable from the tangent to η in C_1 (or C_2).

We find at work here the definition of tangent as the extension of a small side C_1C_2. Thus, C_1C_2 can be considered as a tangent to the curve that intersects AT at point T_2. From T_2, let the perpendicular to AT be traced, such that it meets B_2C_2 and B_1C_1 at points D_2 and D_1, respectively.

By construction, the triangle AC_1C_2 is half of the rectangle with sides B_2D_2 and B_1D_1. Since C_1C_2 is of infinitesimal length, the rectangular strip $B_1D_1D_2B_2$ will also be of infinitesimal thickness, and can therefore be taken as a segment with length equal to the *resecta* T_1C_1. By applying the fundamental idea of the method of transmutation illustrated above, Leibniz could conclude that the ratio between all of the triangular strips concurrent in A, and all of the rectangular strips R_i, with length equal to the *resectae* in C_i is[79]

$$\frac{\sum_{i=1}^{h} T_i}{\sum_{i=1}^{h} R_i} = \frac{1}{2}. \tag{3.11}$$

Since the "slender" triangular strips fill in the surface $A\overset{\frown}{C_i}A$, and the corresponding thin rectangular strips fill in the surface AD_iB_iA, namely, the *figura segmentorum*, it follows,

[78]AVII4, 40, p. 657: "Intelligi poterit constare ex infinitis lineis rectis velut lateribus, quae scilicet portiones sint tangentium, duas applicatas proximas (seu distantia infinite parva a se invicem remotas) iungentium."

[79]See Child (1920, p. 39), Mahnke (1925, pp. 10–11); Hofmann (1974, pp. 54–56).

as a simple consequence of the so-called principle of Cavalieri, that

$$\frac{\overset{\frown}{AC_iA}}{AD_iB_iA} = \frac{1}{2}.$$

The geometric reduction is thus accomplished: the quadrature of a circular segment is reduced to the squaring of the surface bounded by another curvilinear figure, the *figura segmentorum*, which stands in a rational relation with the former. The same reduction can be performed for the other conic sections too, the structure of the reasoning remaining the same.

3.7 Towards the Arithmetical Quadrature of the Circle: The Analytical Solution

In order to express the area of any given sector of the circle, the ellipse and the hyperbola through an infinite series of rational numbers, Leibniz started by calculating the algebraic equation of the *figura segmentorum*.[80]

Let us consider, for simplicity's sake, the case of the circular arc $\overset{\frown}{AC_2A}$, and trace the tangent T_2C_2[81] and the *resecta* AT_2 (Fig. 3.10). Leibniz called the segment AT_2 the "tangent of the arc $\overset{\frown}{AF}$," with F the midpoint of $\overset{\frown}{AC_2}$. The other tangent segment C_2T_2 is then extended to the point O, the foot of the perpendicular drawn from A. Obviously, AO is parallel to the radius BC_2. Leibniz then went on to define two other segments: AB_2, the "*sinus versus*" of the arc $\overset{\frown}{AC_2}$, according to the current terminology borrowed from classical trigonometry, and segment B_2C_2, the "*sinus rectus*" of the same arc.

By virtue of the similitude between triangles BAT_2 and AB_2C_2, the following proportion holds: $AT_2 : AB = AB_2 : B_2C_2$. Since AT_2 and B_2C_2 are parallel, as are also AO and BC_2 (this is by construction), triangles AOT_2 and BB_2C_2 are similar; therefore, a second proportion follows: $AO : B_2C_2 = AT_2 : BC_2$. But it can be easily proven that $AO = AB_2$.[82] Hence: $AB_2 : B_2C_2 = AT_2 : BC_2$, or $AB_2 : B_2C_2 = AT_2 : AB$.

[80]See, in particular, the piece sent to Huygens (and its relative drafts) from 1674: AIII, 1, 39, pp. 142ff.; AVII6, 8, pp. 92ff. See also AVII6, 51, pp. 527–528.

[81]Below, I shall use the term "tangent," consistently with its Seventeenth Century use, to denote a segment cut on the tangent by given points, as in the case at hand, and not the tangent itself, which is a straight line.

[82]In fact, triangles C_2GT_2 and AOT_2 are similar too: the angles OT_2A and GT_2C_2 are equal because they are opposite, whereas angles OAT_2 and GC_2T_2 are both complementary of the equal angles OT_2A and $T_2C_2B_2$, and therefore are equal one with the other. Moreover, we have that $AT_2 = T_2C_2$, so that triangles C_2GT_2 and AOT_2 are congruent too. We have then: $AO = GC_2$ (corresponding sides), and $GC_2 = AB_2$ (segments between parallel lines). By transitivity, $AO = AB_2$.

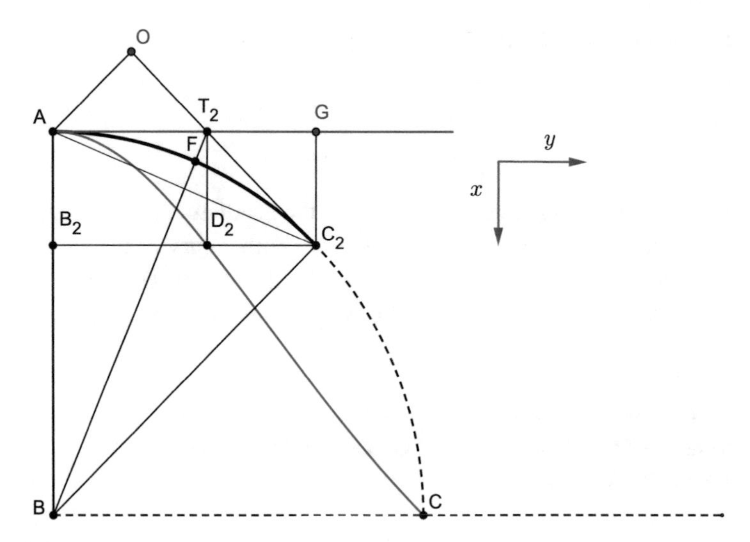

Fig. 3.10 *Figura segmentorum* associated with the circle

Leibniz then proceeded according to the canon of Cartesian analysis, concluding that: $AB_2 = x$, $AB = BC_2 = a$, $AT_2 = y$.[83] From the proportions derived above, he could infer: $B_2C_2 = \sqrt{2ax - x^2}$, and, from $AT_2{:}AB = AB_2 : B_2C_2$ the following:

$$y = \frac{ax}{\sqrt{2ax - x^2}}.$$

In order to eliminate the radical sign, Leibniz rewrote the above expression so as to obtain: $y(\sqrt{2ax - x^2}) = ax$, then raised both members to the square and derived the following equation:

$$x = \frac{2ay^2}{a^2 + y^2}. \tag{3.12}$$

Equation (3.12) defines the curve locus of all points D_i. Since the equation is algebraic, the curve is geometrical in the Cartesian sense. This curve was even left without a proper name in the final draft of *De quadratura arithmetica*. Initially, Leibniz had thought of calling the curve "anonymous," as we can read, for instance, in the draft sent to Huygens in October 1674:

> I truly dare call it "anonymous" by excellence, because even if it has no name, it is one of the most remarkable curve after the conic sections, and much simpler than the Cissoid and the

[83] III, 39, pp. 154, 142.

Conchoid, being only of degree 3, while the conics are of degree 2, and apart from this being of the number of those I call "rational."[84]

The "anonymous curve" is indeed simpler than the conchoid or the cissoid, if we refer, as Leibniz certainly did, to the Cartesian idea of measuring the simplicity of a curve by its degree.[85] On the other hand, Leibniz called the curve "rational," because its associated equation in the form: $x = f(y)$ does not contain any radical expression.[86]

In *De quadratura arithmetica*, Leibniz applied the transmutation method to elliptical and hyperbolic sectors as well, and constructed the corresponding companion curves.[87] Finding the equations of these curves is a matter of simply applying differential calculus (cf. Edwards 1994), although Leibniz, in *De quadratura arithmetica*, as well as in some of its preparatory drafts (for instance, AVII6, 31, p. 365), showed the way in which the same result could be found using Book I of Apollonius's *Conica* and Descartes' finite analysis. To see how this can be done, let us consider a hyperbola and an ellipse with centre E, *latus transversum* AA', and parameter AM.

In Cartesian geometry, we can express the locus of both curves by means of a single equation, employing, as Leibniz did, the ambiguous sign "$\pm$":

$$v^2 = px \pm \frac{px^2}{a}.$$

On the basis of *Conica* I, 13, the following proportion also holds:

$$GA : GA' = AV : VA'.$$

[84]"J'ose bien l'appeler Anonyme par excellence, car quoyqu'elle soit sans nom, elle est pourtant une des plus considérables après les Coniques, et beaucoup plus simple que la Cissoeide ou la Conchoeide, n'estant que de troisiesme degrez, si les Coniques sont du deuxiesme, et outre cela estant du nombre de celles que j'appelle Rationelles" (AIII, 39, pp. 156, 163). In his reply (Cf. AIII, 40, p. 170), Huygens proposed denominating this curve "en luy donnant quelque nom composé des noms de deux lignes dont je trouvois qu'elle estoit produite, qui sont le Cercle et la Cissoide des anciens." Huygens' proposal is clear if we consider the analytical expression of the curve: in fact, each ordinate y can be expressed as: $y = \frac{x^2}{\sqrt{2ax-x^2}} + \sqrt{2ax - x^2} = \frac{ax}{\sqrt{2ax-x^2}}$. With Huygens, we can recognize, in the first term: $\frac{x^2}{\sqrt{2ax-x^2}}$, the expressions of the ordinates of a cyssoid in terms of its abscissae, and in the second one, the expressions of the ordinates of the circle.

[85]cf. Descartes (1897–1913, Vol. 6, pp. 442–443).

[86]I interpret Leibniz's definition in this way: "Figuram rationalem voco cujus abscissae sunt rationales ad ordinatas vel ordinatae ad abscissas, id est quae aequatione exprimi possunt in qua unius incognitae valor purus simplexque est" (AIII, 39, p. 142: "I call that figure the rational figure, whose abscissae are rational to the ordinates, or the ordinates to the abscissas, i.e., which can be expressed through an equation in which the value of one of the unknowns is pure and simple.") See also AIII, 1 73, p. 359.

[87]AVII6, 31; AVII6, 51, Proposition 43.

From the above proportion, we can easily infer $AA' : GA = (AA' \pm AV) : AV$.[88]

Applying the Cartesian equations for the ellipse and the hyperbola in order to express a in terms of the coordinates x and v, and in terms of the parameter p, we arrive at the result that: $GA = \frac{ax}{a \pm x} = \frac{px^2}{v^2}$.[89] In triangle GVC, segments AT and VC are parallel, thus we can easily express the *resecta* AT as a function of the unknowns x and v, namely:

$$AT = y = \frac{px}{v}.$$

We can now eliminate the unknown v in the equation above via the substitution $v = \sqrt{px \pm \frac{px^2}{a}}$ (we consider only the positive root of v) and get

$$y = \frac{px^2}{\sqrt{px \pm \frac{px^2}{a}}},$$

from which we derive, by raising both members to the square and cancelling a factor x from the numerator and the denominator in the fraction on the right,

$$y^2 = \frac{p^2x}{p \pm \frac{px}{a}}$$

and

$$p^2x = y^2(p \pm \frac{px}{a}) = py^2 \pm \frac{pxy}{a}.$$

Collecting the x in the second equation, we obtain

$$x(p^2 \pm \frac{py^2}{a}) = py^2,$$

which yields

$$x = \frac{py^2}{p^2 \pm \frac{py^2}{a}}$$

$$= \frac{ay^2}{ap \pm y^2}.$$

[88] In fact, $GA' = AA' \pm GA$, hence: $GA : (AA' \pm GA) = AV : VA'$. Hence, the previous proportion yields: $(AA' - GA + GA) : GA = (VA' + AV) : AV$ in the case of the ellipse, or $(AA' + GA - GA) : GA = (VA' - AV) : AV$ in the case of the hyperbola. Considering both cases together, we shall have: $AA' : GA = (AA' \pm AV) : AV$.

[89] In the case of the hyperbola: $GA = \frac{ax}{2(a+x)} = \frac{px^2}{v^2}$.

By positing $ap = 1$ (in geometrical terms, the rectangle formed by the *latus rectum* and the *latus traversum* is equal to a unit square: AVII6, Prop. 43, pp. 618–619), we obtain the equation of the *figura segmentorum* for the ellipse and the hyperbola:

$$x = \frac{ay^2}{1 \pm y^2}.$$

We now have all of the elements required to enter the analytical part of Leibniz's quadrature of the circular segment $A\widehat{C_2}A$ (as well as any segment of a central conic section). Leibniz's original procedure, as is very evident in the drafts of *De quadratura arithmetica* composed up to the end of 1675, closely follows Mercator's *Logarithmotechnia* and Wallis's account of this treatise.[90]

Leibniz proceeded in the following way to find the arithmetical quadrature of the segment $A\widehat{C_2}A$ and then of the corresponding circular sector (see Fig. 3.10).[91] He subdivided the given segment $AT_2 = b$ (notice that we now take AT_2 as a known segment with length b) into infinitesimal segments β.[92] Leibniz then computed the area of the trapezoid AT_2D_2A as the aggregate of infinitesimal rectangles with basis equal to β and heights equal to the abscissas x.

Since the equation of the companion curve of the circle, which delimits the trapezoid AT_2D_2A, is $x = \frac{2ay^2}{a^2+y^2}$, the area of half of AT_2D_2A is

$$\frac{AT_2D_2A}{2} = \sum_{i=1}^{i=h} \beta(\frac{a(i\beta)^2}{a^2 + (i\beta)^2}).$$

This sum can be computed applying Mercator's technique and expanding the right-hand member of the above expression into a power series. In short, by positing $a = 1$ and using elementary algebra, the equation of the companion curve to the circle can be rewritten as follows:

$$\frac{x}{2} = \frac{y^2}{1+y^2} = \frac{(y^2)(1-y^2)}{(1+y^2)(1-y^2)},$$

from which we can directly conclude that

$$\frac{y^2}{1+y^2} = \frac{y^2 - y^4}{1 - y^4} = \frac{y^2}{1-y^4} - \frac{y^4}{1-y^4}.$$

[90]The connection between Mercator's and Wallis's quadrature of the hyperbola is less clear if *De quadratura arithmetica* from 1676 is taken into account, because, as we have seen above, this text follows a synthetical approach instead of an analytical one.

[91]AIII1, 39, and AIII1, 72. Cf. also AVII6, 1.

[92]Leibniz defined β as an "infinitesimal part of the radius" (AIII1, 72, p. 344). In AVII6, 1, p. 16, he explicitly refers to Wallis's notation, by indicating the small β as: $\frac{1}{\infty}$.

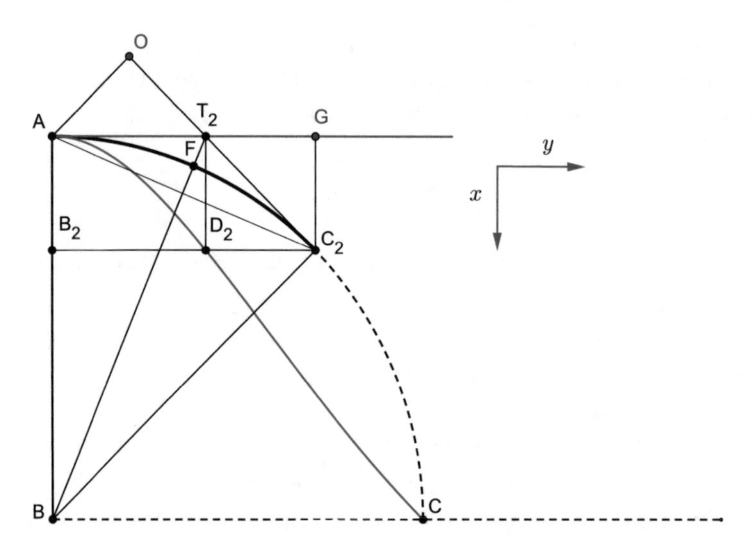

Fig. 3.11 Leibniz's *figura segmentorum* associated with the circle

Expanding both terms $\frac{y^2}{1-y^4}$ and $\frac{y^4}{1-y^4}$ using Mercator's method of long division, Leibniz obtained the following power-series:

$$\frac{x}{2} = \frac{y^2}{1+y^2} = y^2 - y^4 + y^6 \dots$$

On the basis of this result, Leibniz obtained the area of half of the trapezoid $\frac{AT_2D_2A}{2}$ (Fig. 3.11):[93]

$$AT_2D_2A = 2(\frac{b^3}{3} - \frac{b^5}{5} + \frac{b^7}{7} \dots). \tag{3.13}$$

At this point, the area of the segment $\widehat{AC_2}A$, delimited by the arc $\widehat{AC_2}$ and by its subtending chord AC_2 as well as the area of the circular sector ABC_2A, delimited by the same arc $\widehat{AC_2}$ and by radii BA and BC_2 can be easily inferred.

Thus, having posited: $R(AT_2, AB_2)$ as the area of the rectangle delimited by the segments $AT_2 = b$ and $AB_2 = x$, the area of the circular segment $\widehat{AC_2}A$ shall be

$$\widehat{AC_2}A = \frac{xb}{2} - (\frac{b^3}{3} - \frac{b^5}{5} + \frac{b^7}{7} \dots). \tag{3.14}$$

[93] I note that Leibniz's procedure is equivalent to the term-wise integration of the integral: $\int_0^b \frac{y^2}{1+y^2} dy$ (provided $AT_2 = b$), obtained by developing the fraction $\frac{y^2}{1+y^2}$ according to the power expansion illustrated in Eq. (3.7) above.

On the basis of the above equation, Leibniz computed the area of the sector ABC_2A, including that between the arc $\overset{\frown}{AC_2}$ and the radii AB, $BC_2 = 1$, obtaining[94]

$$ABC_2A = b - \frac{b^3}{3} + \frac{b^5}{5} - \frac{b^7}{7} \ldots \tag{3.15}$$

Leibniz derived two important corollaries from Eq. (3.15) above. The first is the rectification of any circular arc, which can be derived, Leibniz noted, by means of "ordinary geometry."[95] In fact, supposing the radius of a given circle being equal to 1, we can obtain, from Eq. (3.15), the length of the arc $\overset{\frown}{AC_2}$ in terms of an infinite series as follows[96]:

$$\overset{\frown}{AC_2} = 2(b - \frac{b^3}{3} + \frac{b^5}{5} - \frac{b^7}{7} \ldots).$$

The second corollary is the arithmetical quadrature of the whole circle. Let us assume this time that the diameter is equal to 1, thus it is sufficient to substitute $b = 1$ in formula (3.15) to obtain the area of the whole circle as a numerical series, namely,[97]

$$A(circle) = \frac{1}{1} - \frac{1}{3} + \frac{1}{5} - \frac{1}{7} + \frac{1}{9} - \frac{1}{11} \ldots \tag{3.16}$$

This is Leibniz's series for $\frac{\pi}{4}$.

---

[94] AIII, 1, 39, pp. 163–164. From Fig. 3.11, $ABC_2A = ABC_2 + \overset{\frown}{AC_2}A$, where ABC_2 is the triangle with side AB and height B_2C_2. By setting: $BA = 1$, $AB_2 = x$, $AT_2 = b = \frac{x}{\sqrt{2x-x^2}}$, $B_2C_2 = \sqrt{2x - x^2}$, and the area of ABC_2 will be equal to: $ABC_2 = \frac{\sqrt{2x-x^2}}{2}$. On the basis of this result, and of the result of (3.15) below, one can derive the area of the sector ABC_2A as: $ABC_2A = \frac{\sqrt{2x-x^2}}{2} + \frac{xb}{2} - (\frac{b^3}{3} - \frac{b^5}{5} + \frac{b^7}{7} \ldots)$. Let us then consider, with Leibniz, the sum: $\frac{\sqrt{2x-x^2}}{2} + \frac{xb}{2}$ (AIII1, 39, p. 164). Since $AT_2 = b = \frac{x}{\sqrt{2x-x^2}}$, we will have: $\frac{\sqrt{2x-x^2}}{2} + \frac{x}{2}(\frac{x}{\sqrt{2x-x^2}}) = \frac{\sqrt{2x-x^2}}{2} + \frac{x^2}{2\sqrt{2x-x^2}} = \frac{x}{\sqrt{2x-x^2}}$. But it has been posited that: $\frac{x}{\sqrt{2x-x^2}} = b$, thus: $\frac{\sqrt{2x-x^2}}{2} + \frac{xb}{2} = b$. Leibniz then concluded: $ABC_2A = b - \frac{b^3}{3} + \frac{b^5}{5} - \frac{b^7}{7} \ldots$. (see Eq. (3.18) in the main text).

[95] AIII1, 72, p. 350, AVII6, 51 , p. 599. Leibniz probably had in mind a few classical results. Thus, in the *Metrica*, Hero quotes the following theorem as a corollary of Archimedes' first proposition of the *Dimensio circuli*: "every sector is half of the rectangle bounded by the periphery of the sector and the radius." The same theorem is evoked in Pappus's Commentary on Ptolemy's Book VI of the *Almagest*. Obviously, a formula for expressing the relation between an arc and its tangent cannot be derived, in the same elementary way, for an arc of hyperbola or ellipse.

[96] See AVII4, 42_2, p. 752; AIII1, 72, p. 345; see also AVII6, 51, Proposition 31.

[97] Cf. AIII1, 72, p. 339; 72, p. 345; 73, p. 356; AVII6, 51, p. 602.

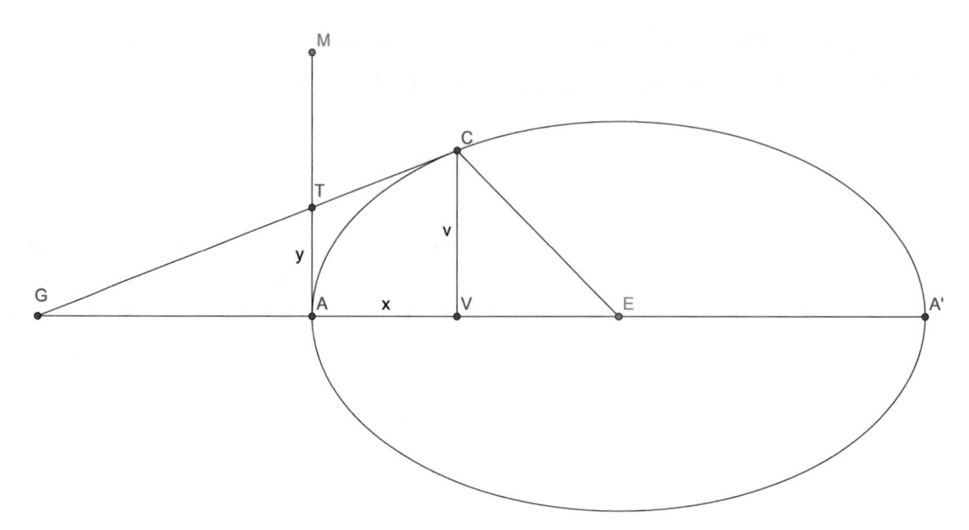

Fig. 3.12 An ellipse with centre E, *latus transversum* $AA' = a$, parameter $MA = p$, $AV = x$, $VC = v$, $AT = y$

By applying the same proof structure, Leibniz was also able to find the area of a sector of an ellipse and a hyperbola and obtained the "general quadrature of the central conic sections." This result comes from a simple application of the method employed to square a circular segment to the ellipse and the hyperbola.[98] Starting from their companion curve with equation:

$$x = \frac{ay^2}{1 \pm y^2},$$

we can, in fact, expand the right-hand member and, applying Mercator's technique of integration, obtain the area of the curvilinear triangle $\overset{\frown}{ATC}$ from point A to point T (see Fig. 3.12; we pose that $AT = b$):

$$\overset{\frown}{ATC} = \frac{a}{2}(\frac{b^3}{3} \pm \frac{b^5}{5} \pm \frac{b^7}{7} \dots) \tag{3.17}$$

The area of the whole sector $\overset{\frown}{ACE}$ can then be derived using elementary geometry, and it will be equal to

$$ACE = \frac{a}{2}(\frac{b}{1} \pm \frac{b^3}{3} \pm \frac{b^5}{5} \pm \frac{b^7}{7} \dots). \tag{3.18}$$

[98] See *De quadratura arithmetica*, Prop. 43.

With this result, Leibniz achieved the arithmetical quadrature of the central conic sections, thus completing the trend of research started by Brouncker and Mercator on the hyperbola.[99] Meanwhile, the use of infinite series represented, for Leibniz, the instrument of a new kind of analysis, a "subsidiary analysis," as he was to put it, which went beyond Viète's and Descartes' finite analysis:

> As Viète and Descartes showed that the Problems of rectilinear Geometry can be reduced to the calculation of numbers by means of equations, I here show how the difficulties of the most important problems of curvilinear geometry can be transferred from Geometry to the arithmetic of rational numbers by means of progressions.[100]

Even if they went beyond the bounds of finite analysis, infinite progressions were, for Leibniz, still anchored in geometrical interpretations. For instance, in *De quadratura arithmetica*, Leibniz reformulated the solutions expressed by the above equations (n. (3.15), (3.16), (3.18)) in the terminology of the theory of proportions, or in terms of geometrical constructions. Thus, Leibniz could claim that the area of a conic sector is equal to a rectangle formed by a side equal to half of the *latus transversum* a and a side equal to the infinite series: $b \pm \frac{b^3}{3} \pm \frac{b^5}{5} \pm \frac{b^7}{7} \ldots$

Leibniz considered that the aforementioned infinite series had a finite sum—provided b is not larger than 1—and thus denoted a segment having finite length.[101] However, Leibniz had doubts about the geometrical character of his solutions. For instance, in concluding one of his drafts on the arithmetical quadrature from October 1674, he remarked that his arithmetical quadrature was not geometrical, since it did not allow one to rectify an arc or to square a sector by means of a construction performable in a finite number of steps.[102] In fact, we could certainly extract from Eq. (3.16) a constructive rule for exhibiting a square equal to a given circle: let us take a square with an area of 1 square foot, and subtract from it a rectangle with an area of $\frac{1}{3}$ square feet. We then add to the resulting figure a rectangle with area $\frac{1}{5}$ square feet, and so on.[103] However, since the series is infinite, truncating it after n finite terms will give us only an approximate geometric construction.

---

[99]Both are recalled in the closing comments added by Leibniz to his draft of *De quadratura arithmetica* from 1674 (AIII1, 39, p. 168).

[100]"Car Viète et Descartes ayant monstré que les Problemes de la Geometrie rectiligne se reduisent au calcul des nombres par les equations, je fais voir icy, comment les difficultez des problemes les plus importants de la Geometrie curviligne sont transferés de la Geometrie a l'arithmetique des nombres rationaux par les progressions" (AIII1, 39, p. 168).

[101]Cf. AIII1, 73, p. 358; AVII6, 51, p. 597.

[102]AVII6, 4, pp. 73–74. This consideration was not a mere passing note, but can be found in the successive drafts of *De quadratura arithmetica* and in the final draft. Cf., for instance, the letter to Gallois (AIII1, p. 356), and AVII6, 51, *Scholium* to Proposition 31, p. 600.

[103]AVII6, 11, p. 111.

Such a use of infinite series in geometric problem-solving created, in Leibniz's mind, a wedge between geometricity and exactness. The classical ideal of a geometrical solution stated that a problem was only truly solved when a construction done in a finite number of steps results in the required object. On the one hand, Leibniz's solution failed to meet this criteria of geometricity. But on the other hand, his arithmetic quadrature was more than just an approximate solution.

Leibniz was indeed particularly adamant in considering his arithmetical quadrature to constitute an exact quadrature. Thus, we read in *De quadratura arithmetica* that the infinite series for $\frac{\pi}{4}$ was constructed according to a law that "impresses the mind," or a law such that the "mind pervades it [namely, the series] all at one blow."[104] It seems that Leibniz's core idea of exactness is hidden behind these suggestive metaphors. In other writings, Leibniz remarked that his alternate series did not merely provide a rule for obtaining increasingly better approximate constructions, but, considered as a whole, expressed *all* of the approximate constructions at the same time.[105] This meant that, simply by relying on the law of formation of the series, one would be able to construct an approximate solution to the circle-squaring problem within any bounds of accuracy as desired, at least in principle.

Nevertheless, one could hope, and Leibniz certainly did at the beginning of his work with problems of quadrature, that an infinite series of rational numbers like the one obtained for $\frac{\pi}{4}$ (or, concerning the hyperbola, $ln(2)$) might be equal to a known rational or radical number, so that the definite quadrature of the central conic sections could be solved by a construction too, exactly as in the case of the quadrature of higher order parabolas and hyperbolas. Eventually, Leibniz's hopes faded away, and the belief that the indefinite and definite geometrical quadratures of the circle were both impossible was to lead him to find a new solution to the disagreement between exactness and geometricity. This topic will be explored in the next sections.

3.8 First Communication of the New Results: Huygens and Oldenburg

Leibniz communicated his discoveries regarding the arithmetical quadrature of the circle to Huygens in summer 1674 (AIII1, 29), sending him the French draft of *De quadratura arithmetica* in the autumn of the same year. Huygens' enthusiastic response reveals that he considered Leibniz's arithmetical quadrature a fruitful path of inquiry that might lead to the discovery of the area of the circle in terms of a rational or radical number, and thus

[104] AVII6, 51, p. 601: "Qua nescio an simplicior dari possit, quaeque mentem afficiat magis. Hactenus appropinquationes tantum proditae sunt, verus autem valor nemini quod sciam visus nec a quoquam aequatione exacta comprehensus est, quam hoc loco damus, licet infinitam, satis tamen cognitam, quoniam simplicissima progressione constantem uno velut ictu mens pervadit."

[105] Cf. AVI, 3, 69, p. 502; AII1, 138, p. 481.

demote Gregory's opposite conviction that the circle-squaring problem could not be solved analytically.[106]

Leibniz initially shared Huygens' optimism, believing that the wedge between geometricity and exactness that he had uncovered could be filled by proving that the sum of the series for the quadrature of the circle was equal to a known species of number. In this case, in fact, the circle-squaring problem would be solvable by a geometric construction, either by ruler and compass or by one of the algebraic curves introduced in Cartesian geometry. Leibniz thus tried to compute the sum of the converging series for the area of the circle, building on his studies on numerical progressions, even if he failed to reach significant advances.[107]

At that time, Leibniz's technique for the computation of series focused on those successions (either finite or infinite) whose terms can be recognized as differences between successive terms of other sequences.[108]

An exemplary case is the finite series: $a_1, a_2, a_3 \ldots a_n$, such that: $a_1 = b_1 - b_2, a_2 = b_2 - b_3, a_n = b_n - b_{n+1}$, where $b_1, \ldots b_{n+1}$ are the terms of a second series, monotonically decreasing.[109]

Leibniz proved that, for any finite sequence $a_1, a_2, a_3 \ldots a_n$, with which we can associate a finite sequence $b_1, b_2 \ldots b_{n+1}$ ($a_1 = b_1 - b_2, a_2 = b_2 - b_3, a_n = b_n - b_{n+1}$), the following result holds:

$$a_1 + a_2 + \ldots a_n = b_1 - b_{n+1}.$$

This elementary theorem states that the sum of the consecutive terms in a difference sequence $a_1, a_2, a_3 \ldots a_n$ is equal to the difference of the first and last term of the base sequence $b_1, b_2 \ldots b_{n+1}$. Leibniz promptly extrapolated this theorem to the case of infinite sequences: this move eventually led him to find, in 1672, the series of the reciprocal of

[106]Huygens expressed his opinion in these terms: "Je vous renvoie, Monsieur, Vostre escrit touchant la Quadrature Arithmetique, que je trouve fort belle et fort heureuse. Et ce n'est pas peu à mon avis d'avoir decouvert, dans un Probleme qui a exercé tant d'esprits, une voye nouvelle qui semble donner quelque esperance de parvenir a sa veritable solution. Car le Cercle, suivant vostre invention estant a son quarré circonscrit comme la suite infinie de fractions $\frac{1}{1} - \frac{1}{3} + \frac{1}{5} - \frac{1}{7} + \frac{1}{9} - \frac{1}{11}$ etc. à l'unitè, il ne paroistra pas impossible de donner la somme de cette progression ni par consequent la quadrature du cercle, apres que vous aurez fait voir que vous avez determinè les sommes de plusieurs autres progressions qui semblent de mesme nature" (AIII1, 40, p. 170). See also Hofmann (1974, p. 82).

[107]Remarkable attempts are contained in the drafts: AVII3, 24, AVII6, 7, p. 90, AVII6, 11, p. 111.

[108]Most of Leibniz's first results on the topic were incorporated into the draft *Accessio ad arithmeticam infinitorum*. This text was prepared by the end of 1672 for publication in the *Journal des Sçavants*. The publication never occurred, so that the *Accessio* remained unpublished during Leibniz's lifetime (AIII1, 2, pp. 2ff.).

[109]See Hofmann (1974, Chapter 2).

triangular numbers, namely: $\sum_{n=1}^{\infty} \frac{2}{n(n+1)} = 2$, as well to treat other series in a similar manner.[110]

Leibniz was probably convinced, in the aftermath of his discovery of the series for the arithmetical quadrature of the circle, that this series or a series extrapolated from it might be obtained as a difference sequence of a still unknown base sequence, and that the limit of $\sum_{n=0}^{\infty} \frac{(-1)^n}{2n+1}$ might be obtained by applying the general theorem to the difference series (or sequences) stated above.

I shall not linger over Leibniz's inquiries into the sum of the series for the quadrature of the circle, since they were soon abandoned.[111] However, it is worth noting that, despite the failed attempts to find the area of the circle, Leibniz remained, at least in the years 1674–1675, confident that his series for the quadrature was the 'royal highway' towards a finite solution of the circle squaring problem.[112] This confirms Hofmann's conjecture that Leibniz's hidden agenda, while studying the quadrature of the circle, was to actually confirm Huygens' belief that π was a rational or a surd number, in any case, a number expressible through known notations.

Meanwhile, Leibniz had also promoted his discovery among mathematicians outside of France. Thus, in July 1674, he described his geometrical discoveries in a letter to Oldenburg, with whom Leibniz was on good terms since his last visit to London:

> I have other theorems, of much greater importance. One of them is admirable in the first instance, by whose aid the area of the circle, or of a given sector of it, can be exactly expressed by an infinite series of rational numbers, continually produced to infinity.[113]

---

[110]Cf. Hofmann (1974, Chapter 2, p. 18), in particular. I point out that the problem of calculating the sum of the series $\sum_{n=1}^{\infty} \frac{2}{n(n+1)}$, suggested by Huygens to Leibniz, had already been solved by Huygens in the 1960s (Huygens 1888–1950, Vol. 14, pp. 50–91). Therefore, it might have been proposed by Huygens as a test in order to assess the ability of the young mathematician. Leibniz found that the reciprocal of the triangular numbers, namely: $\frac{2}{i(i+1)}$ satisfies the following equality: $\frac{2}{i(i+1)} = \frac{2}{i} - \frac{2}{i+1}$, so that, applying the theorem to the difference sequences he found: $1 + \frac{1}{3} + \frac{1}{6} + \ldots + \frac{2}{n(n+1)} = 2 - \frac{2}{n+1}$ and derived the sum of the infinite series of triangular numbers, namely: $1 + \frac{1}{3} + \frac{1}{6} + \ldots = 2$. Leibniz's manipulation of series often led to true results, even if their deduction may not always carry conviction for us (Hofmann 1974, p. 18).

[111]Examples can be found in AVII3, 15, p. 180; AVII3, 38_{10}.

[112]AVII6, 7, p. 91: "Je crois que ceux qui entendent la matiere demeureront d'accord que c'est peut estre le premier Moyen, qu'on ait donné pour arriver à la Quadrature Geometrique du Cercle; et que même la moitié du chemin estant faite, il y a grande apparence, si elle se trouvera jamais, que ce sera par cette voye." See also the *Scholium* (AIII, 1, 39, pp. 165ff.) of the draft for La Roque: "S'il y a lieu d'esperer qu'on pourra jamais arriver à une raison analytique; exprimée en termes finis, du Diamètre à la Circonference, je croys que ce sera par cette voye. Car quoyque les expressions soyent infinies, nous ne laissons pas quelque fois d'en trouver les sommes" (AIII, 1, 72, p. 351).

[113]LSG, I, p. 53: "Alia mihi theoremata sunt, momenti non paulo majoris. Ex quibus illud inprimis mirabile est, cujus ope Area Circuli, vel sectoris ejus dati, exacte exprimi potest per Seriem quandam Numerorum rationalium continue productam in infinitum."

A few months later, Leibniz expounded his arithmetical quadrature to Oldenburg again, remarking:

> No one has given a progression of rational numbers whose sum, continued to infinity, is exactly equal to the circle. It eventually occurred to me, fortunately: in fact I found a very simple series of rational numbers, whose sum equals exactly the circumference of the circle; set the diameter equal to the unity (...) then the proportion of the diameter to the circumference can be exactly exhibited by me via a ratio, not of a number to a number (this would mean to find it absolutely); but via a ratio of a number to a whole series of rational numbers.[114]

Oldenburg was not new to entertaining correspondences on mathematical subjects, but his opinions were often those of his fellows, such as Collins, Pell and Wallis. The latter often sent him letters, to which he gave a more organized shape and translated into Latin.[115] Perhaps also in the case at hand, Oldenburg behaved as a mouthpiece for Collins or Wallis, who both had first-hand knowledge of Gregory's work on the quadrature of the circle and of the subsequent polemics with Huygens, which occurred only a few years earlier.

Oldenburg's reaction towards Leibniz was very tepid. In fact, Oldenburg was unimpressed by Leibniz's solution to the quadrature of the circle, as he avowed that James Gregory had already found a similar result:

> But you truly say that no one has so far given a progression of rational numbers, whose sum, continued to infinity, is exactly equal to the circle (...) but I must add what I have recently received from a man expert in these matters: in fact the aforementioned Gregory is already occupied with such a matter, that he will show in one of his writings that the exactness of the quadrature cannot be obtained.[116]

Leibniz's remark about an exact solution to the circle-squaring problem might have inspired the reference to Gregory's impossibility claim in the excerpt above. As we know from the previous chapter, Gregory had dismissed, in the last pages of the *VCHQ*, any

[114] LSG, I, p. 55: "Nemo tamen dedit progressionem numerorum rationalium, cujus in infinitum continuatae summa sit exacte aequalis Circulo. Id vero mihi tandem feliciter successit, inveni enim Seriem Numerorum rationalium valde simplicem cujus Summa exacte aequantur Circumferentiae Circuli; posito Diametrum esse Unitatem (...) Ratio Diametri ad Circumferentiam, exacte a me exhiberi potest per rationem, non quidem Numeri ad Numerum (id foret absolute invenisse); sed per rationem Numeri ad totam quandam Seriem numerorum Rationalium."

[115] Hall (2002, pp. 160–161).

[116] The letter dates from December 8, 1674: "Quod vero ais, neminem hactenus dedisse progressionem numerorum rationalium cujus in infinitum continuata summa sit exacte aequalis circulo (...) supra dictum nempe Gregorium in eo jam esse, ut scripto probet, exactitudinem illam obtineri non posse" (LSG, p. 57). Oldenburg could be suggesting here that a reissue of the *VCHQ*, which unfortunately has not come down to us, was being prepared around 1673–1674.

search for the exact solution to a problem such as the quadrature of the circle. In geometry, we should rather look for the simplest solution, and, for the circle-squaring problem, this means a solution obtained via an infinite series. On the contrary, by presenting the virtues of his own solution, Leibniz adhered to a methodological criterion of exactness, just like Descartes had done, and attempted to redefine it in order to include his solution based on an infinite series.[117] Therefore, if one understood, as Leibniz certainly did (see the previous section), an "exact" solution to mean a solution that countenances an infinite series obeying certain rules, then Leibniz was certainly justified in claiming that the circle-squaring problem could be exactly solved with the method he provided.

Probably also alerted by Oldenburg's reply, Leibniz began, from 1675, a systematic discussion on the degree of exactness that different solutions to the circle-squaring problem could attain (I will return to this topic in Sect. 3.9 below), and, parallel to this, also a critical study of Gregory's arguments.

3.8.1 Criticism of Gregory

A series of drafts of a letter to Oldenburg written in March 1675 informs us more precisely regarding Leibniz's general dissatisfaction with the theorems of impossibility contained in the *VCHQ*. Even if Leibniz shared Huygens' dissatisfaction with Gregory's proof of his impossibility theorem, he admitted that Huygens' criticism was not persuasive enough and had not closed the question about the analytical or algebraic unsolvability of the circle-squaring problem. As a reaction, Leibniz promised new and original objections that could persuade mathematicians to further investigate the circle-squaring problem:

> Besides the objections made by the celebrated Huygens, for which there is no general consensus, I have peculiar objections too, from which one can adequately conclude that the geometers must not give up this research.[118]

A detailed criticism of Gregory's impossibility arguments can indeed be found in three manuscripts from 1676: *Quadraturae Circuli Arithmeticae Pars Secunda* (AVII6, n. 28, dated June or July 1676), *Series convergentes seu substitutrices* (AVII3, 60, from June 1676), and *Series convergentes duae* (AVII3, 64, June 1676). As the dates reveal, Leibniz

---

[117] See also: Hofmann (1974, p. 100).

[118] AIII1, 46, p. 204: "praeter objectiones ab illustri Hugenio factas, quibus nondum est satisfactum universis, habeo et ego peculiares, unde satis judicari potest, nondum geometras ab hac inquisitione desistere debere". A subsequent letter to Oldenburg, dating from 27 August 1676, shows how Leibniz had not abandon his conviction that Gregory's proofs of impossibility was imperfect and not fully rigorous: "Ceterum ejus demonstrationi editae de impossibilitate quadraturae absolutae circuli et hyperbolae multa haud dubie desunt" (AIII1, 89, p. 580). Analogous remarks can be found in AVII6, 19, p. 175: "Hanc impossibilem esse asseruit ingeniosissimus Gregorius in libro de Vera Circuli Quadratura, sed demonstrationem tunc quidem, ni fallor, non absolvit."

wrote the aforementioned tracts while he was elaborating the ultimate version of *De quadratura arithmetica*, the one including a theorem on the impossibility of squaring a sector of the circle and the other conic sections.

But there is more than a chronological coincidence between Leibniz's criticism of Gregory and the elaboration of the theorem of impossibility appended to *De quadratura arithmetica*. The existence of a connection between Leibniz's own argument and his ongoing criticism of Gregory is especially confirmed by the manuscript AVII6, 28, a draft of *De quadratura Arithmetica* from late Spring 1676. Leibniz concluded it with a *Scholium* containing a long critical discussion of the purported flaws in Gregory's impossibility argument.[119]

Leibniz's account of Gregory's errors begins by making the point about Gregory's strategy (his "*vis argumenti*") for proving the impossibility of analytically squaring a sector of a central conic. Leibniz correctly observed that the gist of Gregory's approach to the quadrature of the central conic sectors consisted in reducing the geometric problem of approximating the area of the sector by polygonal constructions to the problem of computing the limit of a certain convergent sequence (AVII6, 28, p. 352).

As I have surveyed in the previous chapter, Gregory extrapolated, from a couple of polygonal sequences approaching from below and from above a sector of the circle (the ellipse and the hyperbola, respectively) a pair of successions (I_n), (C_n), both monotonically increasing and convergent (cf. *VCHQ*, from Propositions I to VI, pp. 11–19 AVII6, 28, p. 352). It should be pointed out that, in Gregory's terminology, the term 'convergent' has a precise technical meaning, and indicates that the series of the differences $(C_n - I_n)$ becomes smaller than any given quantity as n grows. In other words, the double series (I_n, C_n) is a null series. Moreover, Gregory proved that the pair (I_n, C_n) is defined by the following recursive formula:

$$I_n = \sqrt{I_{n-1}C_{n-1}}, \tag{3.19}$$

$$C_n = \frac{2(I_{n-1}C_{n-1})}{I_{n-1} + \sqrt{I_{n-1}C_{n-1}}}. \tag{3.20}$$

Having obtained this analytical representation of the geometric polygonal construction, Gregory argued, in Prop. XI of the *VCHQ* (*VCHQ*, p. 25), that the limit of the convergent series (I_n, C_n), which expresses the area of the sector, cannot be computed by a finite number of additions, subtractions, multiplications, divisions and root extractions applied to the terms I_n and C_n.

[119]The *Scholium* does not figure in the final version of the *De quadratura arithmetica*. In removing the whole passage, Leibniz was probably obeying precise editorial policy, consisting in separating all historical digressions or philosophical notes from the mathematical content of *De quadratura arithmetica*, and grouping them all together in an introduction, never finished, excerpts from which can be found in: AVII6, 39, 40 49.

In its general outlines, Gregory's argument proved the impossibility of finding an analytical composition f (namely, a finite combination of additions, subtractions, multiplications, divisions and root extractions) such that, applied to any pair (I_n, C_n) and to the sector S, will yield the same quantity K. In symbols: $K = f(I_0, C_0) = f(I_1, C_1 = \dots = f(I_{n-1}, C_{n-1}) = f(I_n, C_n) = f(S, S) \dots$[120] If such a composition could be found, Gregory maintains, S could also be found as the root of the (algebraic) equation: $K = f(S, S)$.[121]

This result is supposed to prove the impossibility of the indefinite quadrature of the circle, since it holds for any sector S that can be approximated by an analytical sequence of polygons. As a corollary, Gregory stated the impossibility of the definite quadrature, that is to say, the quadrature of the whole circle as well.[122]

After a sketchy presentation of Gregory's strategy and main result, the *pars destruens* of Leibniz's considerations properly begins. According to Leibniz, even if Gregory had presented, in the *VCHQ*, an ingenious procedure for computing the limit of convergent series, and thus approximating the area of a conic sector, Gregory's impossibility result was vitiated by a logical flaw ("he somehow sinned in the form of reasoning," AVII28, p. 358).

In Leibniz's view, Gregory had grounded his impossibility proof on the assumption that a convergent sequence tended to an analytical limit only if this limit could be found according to the special method prescribed by Gregory, or that any method capable of computing the limit would be eventually reducible to Gregory's procedure. Since this assumption is by no means evident, Gregory's proof of impossibility as presented in the *VCHQ* was incomplete.

As we have mentioned above, this argument is by no means new, since Huygens and Wallis both levelled the same criticism during their discussions of Gregory's impossibility theorems.[123]

Leibniz's criticism did not stop at this point, however. In fact, in the same tract, AVII6, 28, and particularly in the contemporary manuscript *Series convergentes seu substitutrices* (AVII3, 60), he pushed on with and expanded his critical remarks by explaining Gregory's faulty arguments in light of a mistaken distinction between "formula" and "quantity" (cf AVII3, 60, pp. 758–759). As we read in AVII6, 28:

> It seems to me that I see what has induced into error this very intelligent man, and I have serious reasons to doubt, which would not have displeased Gregory himself if he were still alive. In fact he seems to have reasoned in this way (...) He will say that this is proven [i.e., that it is proven that the sector is not analytical with the sequence of inscribed and

[120] Cf. Lützen (2014, pp. 225–226).

[121] *VCHQ*, XI, pp. 25ff., and Lützen (2014, p. 226).

[122] *VCHQ*, p. 29.

[123] Cf. Huygens (1888–1950, Vol. 3, p. 229). Wallis was especially outspoken in accusing Gregory of having committed a logical mistake: see Wallis (2012, Vol. 3, p. 47).

circumscribed polygons] since we have shown that an analytical formula formed by a and b, in the same way as from $\sqrt{ab}$ and $\frac{2ab}{a+\sqrt{ab}}$ cannot be given. I concede this. But if such a formula, analytically composed, is not given, then an analytical quantity expressed by this formula is not given either. It may be that the quantity is analytical and known, for instance a number; but the formula through which it is composed in the same way from the first pair and from the second pair of terms may be unknown and non-analytical.[124]

It seems to me that Leibniz was addressing a precise criticism, in the passage above, to Gregory's claim about the impossibility of the definite quadrature of the circle. Leibniz illustrated his objection with a simple numerical example: even if the number 3 is analytical both with respect to the numbers 4 and 6 and to the numbers 9 and 13, it could be obtained from the pairs $(4, 6)$ and $(9, 13)$ by means of a non-analytical, or "transcendental," function. One must admit, in fact, that there are examples of transcendental functions, like the logarithms, which can take analytical, i.e., algebraic, values for algebraic arguments.[125]

If we transpose this example to Gregory's result, then it appears that proving the non-existence of an analytical formula for computing the area of a sector of the circle (or of another conic) from the given polygonal series is not sufficient in order to prove that the area of a special sector, like the whole circle, is a non-analytical quantity with respect to the terms of the series.[126]

We can compare Leibniz's objection with the content of Huygens' second reply to that of Gregory, from November 1668, mentioned in the previous chapter (see Chap. 2, Sect. 2.5.1). There are clear similarities between Huygens' remarks and Leibniz's considerations, to the point that one may consider Leibniz's observations as an attempt to

[124]AVII6, 28, p 354: "nam et videre mihi videor, quod in errorem duxerit acutissimum Virum, et rationes dubitandi habeo graves, et ipsi ut arbitror Gregorio si in vivis esset, non displicituras. Itaque sic ille ratiocinatus esse videtur ... Imo vero inquiet, demonstratum est, quoniam ostendimus non posse dari formulam analyticam ex a. et b formatam, eodem modo quo ex $\sqrt{ab}$, $\frac{2ab}{a+\sqrt{ab}}$. Concedo. Si ergo non datur talis formula analytice composita; non datur quantitas analytica per hanc formulam significata. Potest enim fieri ut quantitas sit analytica et nota, verbi gratia numerus; formula autem secundum quam illa eodem modo componitur ex terminis duobus primis quo ex duobus secundis poterit esse ignota et non analytica."

[125]The same problem is discussed in other related tracts. Apart from the AVII6, 28, I refer also to: AVII6, 25, p. 297, and the later texts *De arte characteristica inventoriaque*, from 1678 (AVI4, 78, p. 331) and the *Symbolismus memorabilis calculi algebraici et infinitesimalis*, from 1710. We read in the latter, for instance: "uti impossibilitas extractionis in numeris rationalibus quasitae producit quantitates surdas; ita impossibilitas summationis in quantitatibus Algebraicis quaesitae, producit quantitates transcendentes (...) sane, ut saepe quantitates rationales per modum radicis seu irrationaliter exhibentur, etsi ad formulam rationalem reduci possint; ita saepe quantitates Algebraicae seu ordinariae per modum transcendentium exhibentur, etsi eas ad formulas ordinarias reducere liceat" (Leibniz 2011, p. 275).

[126]AVII3, 60, p. 759, AVII6, 28, p. 354.

make Huygens' original objection more precise, and therefore more persuasive, thanks to the conceptual distinction between quantities and formulas.

Thus, aware of the distinction between analytical and non-analytical formulas and quantities, which, in his opinion, tainted Gregory's argument, Leibniz opted for a "wholly new approach" to the proof of the impossibility of squaring a circular, elliptical and hyperbolic sector.[127] His new strategy is simple: whereas Gregory set out to solve the problem of determining the area of a sector from a given formula produced by one and the same operation, and ended up with its impossibility, Leibniz set out to solve the problem of determining the relation between the area of a sector and its tangent, and proved the non-analytic, or transcendental, nature of this relation.[128] As we would say today, Leibniz's result amounts to proving the non-algebraic nature of certain functions (namely, the trigonometric functions sin or arcsin, and the logarithmic function, for that which concerns the hyperbola).

In a note written between April and June 1676, titled *Impossibilitas quadraturae circuli universalis*, Leibniz further clarified the meaning of the circle-squaring problem (and of its relative impossibility) in the following terms:

> The quadrature problem is twofold: there is a universal and a particular quadrature. The universal quadrature exhibits a rule with whose aid any portion of the circle can be measured, or with whose aid, from a given tangent (or sine) the arc or the angle can be found. And then there is the particular quadrature, which exhibits a certain part of the circumference (and those sectors, whose ratio with that part is known). Hence, if one exhibited the whole circle or the whole circumference, and nothing but these sectors whose ratio with the circumference is already known, one would not thereby achieve the desired universal quadrature.[129]

By distinguishing "universal" and "particular" quadratures, Leibniz was perhaps among the first mathematicians to render explicitly and precisely, on a terminological level, the customary distinction, from the second half of the Seventeenth Century onwards, between two sorts of problem related to the quadrature of a curve.[130] On the one hand, universal quadratures determined the area included between the curve and two arbitrary coordinates.

[127] AVII3, 60, p. 758.

[128] AVII3, 60, p. 758.

[129] AVII6, n. 18, p. 165: "Quadratura duplex est, universalis et particularis: Universalis, quae regulam exhibet cujus ope quaelibet Circuli portio possit mensurari, seu cujus ope ex data tangente (vel sinu) possit inveniri arcus sive angulus. Particularis , quae certam circumferentiae portionem, (: et eas, quarum ad hanc portionem nota est ratio:) exhibet. Unde et si quis totum circulum totamve circumferentiam exhiberet, non vero nisi eas partes, quarum ad circumferentiam nota jam tum est ratio, is quadraturam, qualis desideratur, Universalem non dedisset."

[130] The term "universal quadrature" was previously used by Mengoli to refer to Archimedes' quadrature of the parabola. Cf. Mengoli (1650), *preface*: "Meditanti mihi persaepe Archimedis parabola Quadraturam, propterquam infinita triangula in continu? quadrupla proportione existentia certos limites quantitatis non excedunt; occurrit universalis illa Quadratura eiusdem argumenti occasione a Geometris demostrata, qua magnitudines infinita continuam quamlibet proportionem maioris

On the other hand, particular quadratures determined the area of the whole figure. In the case of the circle, determination of the areas of an arbitrary circular sector is an example of the former, and the traditional circle-squaring problem is of the latter kind.[131]

As the title of AVII18, *Impossibilitas quadraturae circuli universalis*, makes clear, Leibniz claimed the impossibility of the universal quadrature of the central conic sections, by which he meant the impossibility of finding an algebraic relation between the arc and its corresponding tangent.

On the other hand, he maintained that the question of the possibility or impossibility of the particular quadrature—that is to say, the question of whether the circle might be analytical with respect to, or even commensurable with, the square constructed on its diameter—was not a question that had yet been settled.[132] This opinion persists in *De quadratura arithmetica*, in which the closing proposition only refers to the universal, or general, quadrature of the circle and of the other central conic sections.

3.9 The Classification of Quadratures and the Impossibility of Squaring the Circle

Aside from sparse notes from 1674 and 1675 (AVII3, 39, p. 589; AVII5, 26, p. 203), most of Leibniz's considerations on the impossibility of squaring a central conic section date back to 1676, when they appear in a number of manuscripts related to the quadrature of the circle (AVII6, 18, p. 166, AVII6, 19, p. 176, AVII6, 28, pp. 350ff., AVII3, 60, pp. 758ff.), and in a more complete form in Proposition 51 of *De quadratura arithmetica*. We read there:

> It is impossible to find a better general quadrature of the circle, the ellipse or the hyperbola, or a relation between the arc and its chords, or between the number and its logarithm, which is more geometrical than our own. This proposition stands as the crowning of our theory.[133]

inaequalitatis possidentes in praefinitas homogeneas quantitates colliguntur." On the intellectual relations between Mengoli and Leibniz, see Massa (2016).

[131]It should be pointed out that Leibniz did not strictly adhere to his own terminology, and sometimes employed the term "general" as a synonym for "universal." A notable case is AVII6, 51, Prop. 51, as I expound below.

[132]Regarding this concern, Leibniz affirmed, in AVII6, 18: "Certas autem partes vel etiam totum Circulum (: sed non quamlibet ejus portionem:) analytice inveniri posse, nondum despero." ("I have not lost the hope yet that precise parts [*certas autem partes*] or even the whole circle (but not any of its portions) can be found out analytically.")

[133]AVII6, 51, p. 674 (or LQK, p. 134): "Impossibile est meliorem invenire Quadraturam Circuli Ellipseos aut Hyperbolae generalem, sive relationem inter arcum et latera, numerumve et Logarithmum; quae magis geometrica sit, quam haec nostra est. Haec propositio velut coronis erit

Leibniz's proposition is structured by an odd grammar that makes appeal not just to one but to two comparative forms: it is impossible—Leibniz wrote—to find a better quadrature ("*meliorem quadraturam*"), of the circle and the hyperbola, or a more geometrical relation ("*relationem quae magis geometrica sit*") than the one presented in the treatise (and summarized above in Sect. 3.1).

But in what sense might one solution to a quadrature problem be said to be better than another one? As regards the case of the circle (although the same reasoning can hold for the case of the ellipse and the hyperbola), an answer can be advanced considering Leibniz's attempt, consigned to a draft of the already mentioned letter to Oldenburg from 1675, to establish a hierarchy among several types of solution to the universal or particular circle-squaring problem according to their exactness.

Leibniz further refined this classification in a tract from 1676: *Praefatio opusculi de quadratura circuli arithmetica* (AVII6 19, pp. 176–177), a purported preface, as its title indicates, to a contemporary draft of *De quadratura arithmetica*, now published in AVII6, 20.[134]

The classification presented in the *Praefatio*, the most elaborate of the two, follows both a conceptual and a historical rationale (Fig. 3.13). Leibniz proceeds, in fact, by classifying, through successive dichotomies, actual solutions to the circle-squaring problem and merely conceivable ones. The first, most general subdivision demarcates quadratures obtained by tentative, empirical procedures, such as those obtained by unwinding a string around the circumference of a material circle, from "rational quadratures," which involve solutions obtained through proper mathematical methods ("*arte ... et secundum regulam*"). Leaving "empirical" quadratures aside, Leibniz went on to clarify the variety of rational quadratures. Firstly, he distinguished two sub-species: "approximate" and "exact" quadratures. Both of them could be obtained either by geometric constructions ("*per ductum linearum*", AVII6, 19, p. 172) or by a numerical solution ("*per calculum*", AVII6, 19, p. 172. See also Knobloch 1989, p. 130).

Among "approximate quadratures," Leibniz counted both geometric constructions gauging the circle in the style of the Archimedean polygonal method and computations that

contemplationis hujus nostrae." As has been suggested by the editors of AVII6, Leibniz employs a similar construction in a letter to Oldenburg from August 1676, in which we read: "Non credimus, meliorem circuli quadraturam linearem quam haec est unquam datum iri" (AVII6, 51 p. 520). For that which concerns the relation between numbers and their logarithms, on the one hand, and the quadrature of conic sections, on the other hand, suffice to say that Leibniz argued for the impossibility of finding an algebraic universal quadrature of the hyperbola on the grounds of the connection between the hyperbolic areas of an equilateral hyperbola and the logarithmic function. See Sect. 3.5, and below, Sect. 3.10.

[134] I point out that the *Praefatio* is not a preface intended for *De quadratura arithmetica* (AVII6, 51). Interestingly, quadratures are classified along similar lines in the later article *De vera proportione circuli ad quadratum in numeris rationalibus expressa*, published in 1682 (LSG5, p. 120).

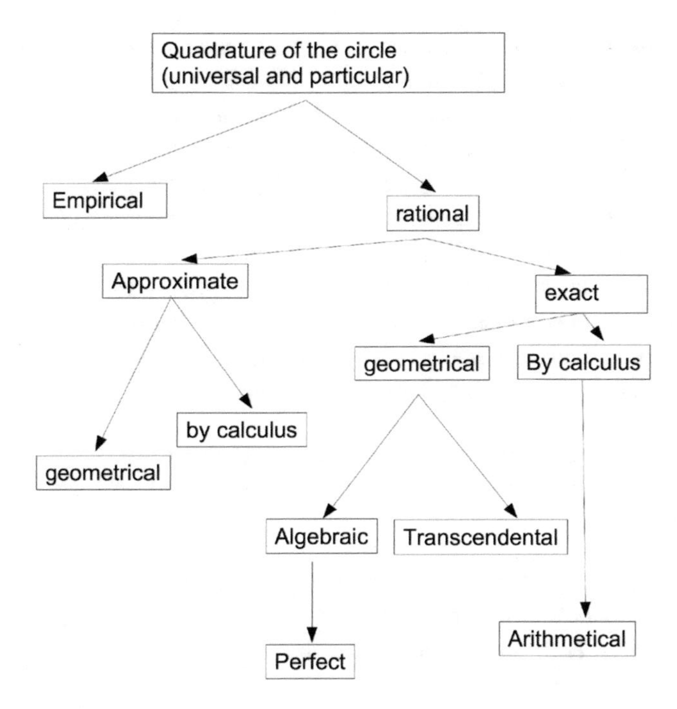

Fig. 3.13 Leibniz's classification of quadratures. The figure is taken from Knobloch (1989, p. 130), with minor changes

express π by irrational estimates (in Leibniz's terminology, these approximate quadratures are obtained "by a calculus.")[135]

By contrast, an "exact geometric" quadrature amounted, according to Leibniz's scheme, to a solution obtained by intersecting curves, either algebraic or transcendental. By virtue of the equivalence between geometrical and algebraic curves in Descartes' geometry,[136] a geometric solution obtained by algebraic curves would stand on a par with an "analytical" quadrature, which expresses the area of the circle or of the sector as an algebraic function of the sine or the tangent and of the radius.

Leibniz acknowledged that solutions expressing the area of the circle (or of any circular sector) by a quantity or a sequence of quantities "whose nature and rule of continuation

[135]Geometric and numerical quadratures were often intertwined. Among early modern representatives of approximate quadratures, Leibniz names Metius, Snell and Van Ceulen, three renowned Renaissance mathematicians who had worked on finding better approximate values for π, and the contemporary mathematician James Gregory. Even if he does not mention him in the *Praefatio*, Leibniz probably also had Huygens in mind (AVII6, 2).

[136]On the equivalence between the two types of curve in Descartes' geometry, see Bos (2001, p. 336).

is known" should also be considered exact.[137] In particular, the solution presented in *De quadratura arithmetica* belongs to this category (AVII6, 19, p. 175).

Leibniz conceded that the arithmetical quadrature was not the most exact conceivable type of solution for the universal or particular circle-squaring problem, since one could certainly conceive (and some even tried to realize) the "analytical" or "geometric" quadrature of the circle and of all its sectors as the most exact or "perfect" quadrature, insofar as it does not make appeal to infinite expressions.[138]

However, the concept of a perfect quadrature is explicitly ruled out by Leibniz as contradictory: "it is impossible," Leibniz stated in the *Praefatio*, "to express the general relation between a circular arc and its sine by an equation of a certain dimension."[139] From this assertion, Leibniz immediately derived the corollary:

> No full, analytical quadrature can be found which, while being expressed by an equation the values of the terms of which are rational numbers, would be more perfect than the one which we have given.[140]

As we have seen, the same theorem is presented again with only minor variations, excepting a generalisation to the squaring of the hyperbola, in Proposition 51 of *De quadratura arithmetica*, probably appended to the treatise in 1676.[141]

3.9.1 An Impossibility Proof

The earliest known proof that a perfect quadrature of the circle is impossible can already be found in the *Praefatio*. As regards its structure and content, the proof is very similar to the argument given in AVII51, which has recently been studied by Lützen (2014, p. 233). In order to integrate the account given by Lützen into the present inquiry, I shall present here the version of the impossibility proof given in the *Praefatio*, which can be considered the earliest argument elaborated by Leibniz.

As in *De quadratura arithmetica*, Leibniz reasoned by contradiction, and assumed that there exists an algebraic equation in the form: $P(\sin(v), v) = 0$, where P is a

---

[137] AVII6, 19, p. 174: "Valor exprimi potest exacte, vel per quantitatem, vel per progressionem quantitatum cujus natura et continunandi modus cognoscitur."

[138] AVII6, 19, p. 175: "Perfecta autem Quadratura illa erit quae simul sit Analytica et linearis, sive quae lineis aequabilibus, ad certarum dimensionum aequationes revocabilibus, construatur."

[139] AVII6, 19, p. 175: "Sed relationem arcus ad sinum in universum aequatione certae dimensionis explicari impossibile est."

[140] AVII6, p. 176: "Quadraturam plenam, analyticam, aequatione expressam, cujus terminorum dimensiones sint numeri rationales, perfectiorem, quam dedimus . . . reperiri non posse."

[141] See AVII6, 28, p. 348, for an intermediate version from June or July 1676. This version does not contain any reference to the impossibility of squaring the hyperbola.

polynomial of finite degree m, expressing the relation between a circular arc v and its sine (AVII6, 19, p. 175). Therefore, the roots of the equation $P(\sin(v), v) = 0$ can be constructed, according to the Cartesian canon for the construction of equations, by intersecting algebraic curves.[142] The easiest way to perform this construction is by intersecting the curve associated with the (algebraic) equation $P(\sin(v), v) = 0$ with a straight line.

As Leibniz explained, a simple way to construct a curve associated with the equation $P(\sin(v), v) = 0$ requires a pointwise construction, obtained by applying, ordinatewise, each sine to successive arc-lengths.[143]

A construction of this curve, called *linea sinuum* or *curva sinuum*, is given in Proposition 48 of *De quadratura arithmetica*, according to the following procedure.[144]

Let the circular arc EFR be given (see Fig. 3.14), wih radius ED and center D. Let an arc EF be taken on EFR, and let us take, or suppose given, a segment DB on DA, such that $DB = arc(EF)$ (notice that the construction of the curve of the sines requires a procedure for rectifying any arc of the circumference). From B, let us trace a segment BC, orthogonal to AB, and equal to the sine FH of the arc EF. If we repeat the same construction for any other arc, we will determine a collection of points: $C, C_1, C_2 \ldots$, each one corresponding to arcs $EF, EF_1, EF_2 \ldots$. The *linea sinuum* will be the locus of points C_n.[145]

Moreover, since the curve is supposed to be algebraic by virtue of the *reductio* assumption, the curve is receivable in Cartesian geometry.[146]

---

[142] For an overview and discussion on the history of the Cartesian technique for the construction of equations, see: Bos (1984). Leibniz was certainly familiar with this technique, and he had made interesting contributions himself (as in *De constructione*, AVI 3, 45).

[143] AVII6, 19, p. 175: "Hoc posito linea curva ejusdem gradus delineari poterit, ita ut abscissa exprimente sinus, ordinata exprimat arcus, vel contra. Hujus ergo lineae ope poterit arcus, vel angulus in data ratione secari, sive arcus, qui ad datum rationem habeat datam, inveniri sinus ..."

[144] Leibniz probably came to know this curve from Honoré Fabri's treatise *Opusculum geometricum de linea sinuum et cycloide*, published in 1659. Cf. Fabri (1659, pp. 5 and 10).

[145] AVII6, 51, p. 642. The procedure explained by Leibniz corresponds, in a more modern guise, to the plotting in a Cartesian reference frame of an arbitrary number of points whose abscissas correspond to the sines of given arcs, and whose ordinates express the corresponding arc-lengths: since, in the setting of Seventeenth Century geometry, one coordinate was not preferred to another, the curve thus obtained might be identified with the graph of the arcsin function.

[146] Relying on Descartes' *Géométrie*, in fact, Leibniz accepted the alleged equipollence between the expressability of a curve through an algebraic equation and its constructibility by a system of "rulers and compasses intertwined, that push and guide each other" (AIII1, 46, p. 204), namely, articulated devices possessing one degree of freedom, so as to assure the unicity and continuity of the tracing motion. See also Descartes (1897–1913, Vol. 6, pp. 391–392).

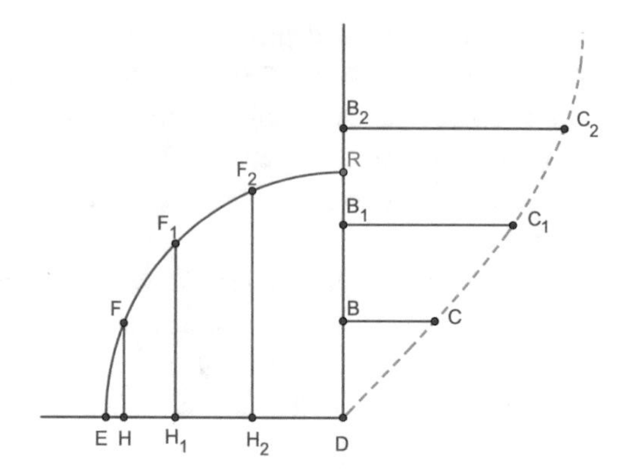

Fig. 3.14 A pointwise construction of the curve of the sines (Cf. *De quadratura arithmetica*, 48)

The *linea sinuum*, as was promptly noted by Leibniz, can be successfully employed to divide any given arc not greater than a quadrant into n equal parts. For instance, if we want to divide the arc EF (Fig. 3.14) into n equal parts by means of this curve, it will be sufficient to trace its sine FH, and construct a segment $DK = FH$ on the extension of the radius ED. Let the perpendicular to DK then be constructed on point K, and let C be the intersection point between the perpendicular and the curve of the sines (which are supposed to be traced): the normal to C on DA will cut this axis at a point B. We shall then have $DB = arc(EF)$, so that the arc EF is rectified. In order to solve the initial problem, it is sufficient to divide, by euclidean means, the segment DB into n parts, and find the corresponding sines.

Since this problem has been solved solely through the use of supposedly algebraic curves, the problem of dividing an angle into an arbitrary number of parts is algebraically solvable too. But this is absurd, Leibniz insisted, because:

> It is well-known indeed that there are so many various degrees of the problems, as many as the (at least odd) numbers of the sections. Indeed, bisecting an angle is a plane problem, trisecting is a solid or conic problem, dividing the angle into five parts a supersolid problem, and so on indefinitely. The higher the problem, the greater the number of equal parts in which the angle must be divided. This is admitted by the Analyticians, and it could be proved universally, if we had space. Thus, it is impossible to express the relation between arc and sine universally, with a single equation of determinate degree.[147]

[147] AVII6, 19, pp. 175–176: "Constat enim tot esse varios gradus problematum, quot sunt numeri (saltem impares) sectionum; nam bisectio anguli est problema planum, trisectio problema solidum sive Conicum, quinquesectio est problema surdesolidum, et ita porro in infinitum, altius problema prout major est numerus partium aequalium, in quas dividendus est angulus; quod apud Analyticos in confesso est, et facile probari posset universaliter, si locus pateretur. Impossibile est ergo relationem arcus ad sinum, in universum certa aequatione determinati gradus exprimi."

In the passage above, Leibniz referred to a result on the theory of angular sections to be found in François Viète's posthumous work, *Ad angularium sectionum analyticen theoremata καθολικώτερα*.[148]

Viète's treatise deals with what we might call, in modern mathematical terminology, the study of the trigonometric functions of the multiples of a given circular arc or angle. The presumed aim in his treatise *Ad angularium sectionum* was to give an algebraic treatment of the relations between trigonometric lines (sines and cosines) associated with arcs and angles. As a result, he consequently tabulated the coefficients of equations expressing the relations between the sine of an angle v and the sine of its submultiples $\frac{v}{n}$, for several n. The schema of coefficients, constructed according to a recursive rule, was to enable him to extrapolate the equations corresponding to the division of the angle into any number n of parts (where n is an integer). In this way, Viète claimed to have given the analytical translation of the more general problem of finding: "one angle to another as one number is to another," namely, the problem of the general section of the angle.

One of Viète's main results obtained in the treatise on angular sections can be thus resumed, with the aid of a more expedient formalism than Viète's original symbolism:

$$sinv = \sum_{k=0}^{n} \binom{v}{n} cos^k(\frac{v}{n}) sin^{n-k}(\frac{v}{n}) sin(\frac{1}{2}(n-k)\pi).$$

As Leibniz remarked, the analytic treatment of the angular section problem involved the following corollary (not pointed out by Viète, by the way): following the procedure described in the *Sectiones angulares*, each instance of the problem of dividing an arbitrary angle into n parts can be associated with an equation of n-th degree at most. This fact can be gleaned through the examination of the equation above, relating $sin(v)$ to $sin(\frac{v}{n})$: at least, if $n = 2k+1$, the odd powers of cos^k in the previous expression will be equal to zero, and the even powers can be rewritten so as to eliminate the cosines thanks to trigonometric equalities.[149]

On the basis of this result, Leibniz concluded that the problem of the general angular section could not be associated with a single polynomial equation in a finite degree. But this very problem is solved, for any n, by the *linea sinuum* (inasmuch as, for every n, this curve constructs, at most, through the intersection with a straight line, the n -section of a given angle). Thus, it is not possible to associate this curve with a polynomial equation in a finite, determinate degree either.

There thereby arises a contradiction, from which it follows that: "it is impossible to express the relation between arc and sine universally, with a single equation of determinate

[148]The treatise was first published in 1615, with some additions by Alexander Anderson, and it was reprinted in 1646 with a slightly different title, in the edition of Viète's works edited by Frans van Schooten (cf. Viète 1646, pp. 286–304). See also: Viete (1683, pp. 418–450).

[149]See Lützen (2014, pp. 235, 236).

degree."[150] As a consequence, not only could Leibniz conclude the impossibility of the universal quadrature of the circle, since the relation between an arc and its corresponding sine cannot be expressed by a final algebraic equation, but he was also able to establish the transcendental nature of a curve, namely, the *linea sinuum*.

The proof of the same theorem given in *De quadratura arithmetica* follows an analogous structure, save for the dismissal of the curve of the sines, which is not strictly necessary for concluding the *reductio* argument, and for a reference to the division of the angle into a prime number instead of an odd number of equal parts.[151] This small but significant change is probably a consequence of the elementary fact that the problem of dividing an angle into an odd, non-prime number m of equal sectors can be further reduced to the more elementary problem of dividing the angle into each prime factor of m. For instance, solving the problem of dividing an angle into 15 equal parts amounts to solving the problem of dividing the original angle into five parts, and then dividing each angle equal to one fifth of the original one into three equal parts.

Leibniz took for granted that equations corresponding to divisions into a prime number of parts were irreducible to lower degree equations. From our viewpoint, this claim, crucial for the completion of the impossibility argument, is by no means obvious and also needs to be adequately proved.

However, it can be supposed that early modern and later mathematicians assumed Viète's insight into the algebraic structure of the angular section problem to be correct and definitive. Two historical clues can add plausibility to this supposition.

A first clue in support of this possibility can be found in a Seventeenth Century treatise in analytic geometry, namely, the *Analytic treatise on the conic sections*, written by G. de l'Hôpital (1661–1704) and published posthumously in 1707 (see Hopital 1707). In this text, the problem of the angular division is taken into account in connection with a discussion about the construction of regular polygons. Building directly on Viète's results, De l'Hôpital asserted without proof that the equations derived by Viète in order to deal with the problem of analytically dividing the angle into an arbitrary number of parts were: "the simplest ones, when the number of equal parts is a prime number."[152]

A second clue can be found in a later work, the *Histoire des recherches sur la Quadrature du cercle*, written by Jean Etienne Montucla (1725–1799) and originally published in 1758. Although this text appeared well into the Eighteenth Century, it is related to our topic, since it contains one of the first published presentations of an argument for the unsolvability of the universal circle-squaring problem that sums up, in its essential outlines, Leibniz's argument from *De quadratura arithmetica*, without, however, mentioning it. The argument related by Montucla takes the form of a *reductio*,

[150] AVII6, 19, p. 176.

[151] For a reconstruction of this proof, see Lützen (2014), in particular, pp. 234–236.

[152] "Les plus simples qu'il est possible, lorsque le nombre des parties égales est un nombre premier" (Hopital 1707, p. 418).

and begins from the assumption that a circular sector or arc is a finite algebraic function of its corresponding cosine and, conversely, the cosine is a finite algebraic function of the corresponding arc.[153] Therefore, Montucla argued, the problem of finding the ratio between two arcs will be expressed by a finite algebraic equation too. In this way, the circumference can always be divided into two parts according to any given ratio between two integers resorting to a finite algebraic equation with a determinate degree n. But this is impossible, Montucla concluded, on the grounds of the theory of angular sections, since: "if the ratio between two arcs AB and AE, is expressed by two numbers, prime with respect to one another, and greater than n, the resulting equation will necessarily be of a degree higher than n."[154]

Not only did Montucla simply rephrase, in the language of Eighteenth Century analysis, the same argument as Leibniz, he did not add anything to Leibniz's argument as regards the reducibilty of the equations associated with angular sections either. On this subject, he merely remarked: "whatever may be the number n, it cannot be finite and determined, since it must respond to all the imaginable cases of angular sections, and there is an infinity of them leading to equations of an infinite degree."[155]

One may suppose that, had he or his contemporaries found any flaw in the use of angular sections to prove the impossibility of the universal quadrature of the circle, such a flaw would have been made explicit at some point in Montucla's account.

For an explicit criticism of Montucla's proof, we have to wait until the second edition of Montucla's *Histoire*, which appeared in 1831 with an *addendum* to the proof of the impossibility of solving the universal quadrature of the circle by algebraic means (Montucla 1831, p. 110). Since the second edition appeared posthumously, that addition was probably made by the editor S. F. Lacroix (1765–1843).

In particular, the author of the *addendum* observed that Montucla's reasoning could not yet be taken as a proof, although the numerous, failed attempts to solve the circle-squaring problem made its belief in the unsolvability by algebraic methods quite plausible. This remark was followed by a summary of Lambert's proof of the irrationality of π (a problem not tackled by Leibniz in *De quadratura arithmetica*).

Although concise, the editor's note is telling. Since it was added for the second edition of 1831, it shows that, by that time, Montucla's argument (and Leibniz's argument *a fortiori*) were viewed as questionable. Unfortunately, the editor did not give details as to why exactly that impossibility proof was no longer judged to be convincing.

[153]Montucla (1831, pp. 108–109).

[154]Montucla (1831, p. 109): "car la raison proposée entre les arcs AB, AE, étant exprimée par deux nombres premiers entre eux et plus grands que n, l'équation qui en résultera sera nécessairement d'un degré plus élevé que *n*."

[155]Montucla (1831, p. 110): "Quel que soit le nombre n, il ne peut donc être fini et déterminé, puisqu'il doit répondre à tous les cas imaginables des sections angulaires, et qu'il y en a une infinité qui conduisent à des équations d'un degré infini."

Meanwhile, let us note that the impossibility of squaring the whole circle algebraically entails the impossibility of the universal quadrature of the circle. The reference to Lambert in the second edition of Montucla's book might suggest that mathematicians had, by the beginning of the Nineteenth Century, turned their attention to the problem of determining the nature of π, which is indeed sufficient to settle the question about the universal quadrature of the circle too. Eventually, the proof that π is a transcendental number, provided by Lindemann in 1882, put a final word to the questions about the possibility of the universal and particular quadratures of the circle.

It was probably as a consequence of these changes in the mathematical agenda that, in the course of the Nineteenth Century, Montucla's and Leibniz's impossibility arguments would gradually glide towards the background and be forgotten, becoming not only unconvincing, but also superfluous.

3.10 The Impossibility of Finding the Universal Quadrature of the Hyperbola

Leibniz gave a similar impossibility proof for the universal quadrature of the hyperbola in the final section of Proposition 51:

> In the same way, once a general relation between arcs and cords has been found, the universal section of the angle could be given by one equation of determinate degree; so once a general quadrature of the hyperbola, namely a relation between a number and its logarithm, has been found, any number of mean proportions could be found by one equation of determinate degree; which is absurd, as mathematicians know (...) thus it is impossible to find a general quadrature, or a construction applying to any given sector of the hyperbola, or of the circle and the ellipse, which is more geometrical than our own.[156]

The proof for the impossibility of squaring the hyperbola given in *De quadratura arithmetica* follows the same pattern as the proof concerning the impossibility of the universal quadrature of the circle. In fact, just as the latter relied on the alleged impossibility of expressing the problem of dividing an angle into an increasing (prime) number of equal parts via an explicit algebraic equation of a fixed, finite degree, so the former relies on the analogous impossibility of expressing, via an analogous algebraic equation, the problem of inserting an arbitrary number of mean proportionals between two given quantities.

[156]AVII6, n. 51, p. 674: "nam, quemadmodum generali relatione inter arcum et latera inventa posset haberi sectio anguli universalis, per unam aequationem certi gradus; ita generali inventa quadratura hyperbolae sive relatione inter numerum et logarithmum, possent inveniri quotcunque mediae proportionales ope unius aequationis certi gradus, quod etiam absurdum esse, analyticis constat (...) Impossibilis est ergo quadratura generalis sive constructio serviens pro data qualibet parte Hyperbolae aut Circuli adeoque et Ellipseos, quae magis geometrica sit, quam nostra."

Logarithms are useful for solving the problem of inserting an arbitrary number of mean proportionals, in the following way. In fact, inserting an arbitrary number of mean proportionals between two given quantities is equivalent to the construction of a certain geometric progression. For example, let us suppose that we want to search for n (with n being any natural number) mean proportionals between the given quantities a and b. To solve this problem, it is sufficient to construct the geometric progression: $a, ar, ar^2; \ldots; ar^{n+1}$, with $r = (\frac{b}{a})^{\frac{1}{n+1}}$. A simple application of logarithms yields the following equation:

$$log(x) = log(a(\frac{b}{a})^{\frac{1}{n+1}}).$$

This equation can be further simplified, by applying a few elementary manipulations:

$$log(x) = log(a) + \frac{1}{n+1}(log(b) - log(a)).$$

Therefore, once $log(a)$ and $log(b)$ are known, for example, by means of a table, finding $log(x)$ is a matter of simple arithmetic computations. Once the corresponding quantity x is found, the problem of inserting n mean proportionals is solved too.

Given these premises, Leibniz's impossibility proof becomes straightforward. On the basis of the relation between the hyperbola-area and the logarithms, if the relation between the area of any sector of a hyperbola and its basis (we can think, for simplicity's sake, of an equilateral hyperbola) could be expressed algebraically, then the relation between a number and its logarithm could be expressed algebraically as well. But this conclusion is absurd. In fact, by virtue of the relation between the problem of constructing mean proportionals and the problem of computing the logarithm of a number, illustrated above, if the relation between a number and its logarithm were expressed by a single, explicit algebraic relation, then the problem of inserting an arbitrary number n of mean proportionals between two given segments a and b or, equivalently, the problem of dividing the ratio between a and b into $n + 1$ parts, would be algebraically solvable too.

However, just as in the case of the division of the angle, Leibniz assumed that the equations associated with every instance of the problem of inserting a prime number m of mean proportionals increases with m. Therefore, contrary to our assumption, the general problem of inserting m mean proportionals between two given segments cannot be expressed through an algebraic equation of fixed degree (AVII6, 51, p. 556), and the impossibility of solving the quadrature of the hyperbola algebraically is therefore proved.

3.11 The Significance of Leibniz's Impossibility Result

In the historical setting of Seventeenth Century geometry, the significance of the impossibility result proved by Leibniz is, at first sight, not obvious, since it seems to be at odds with respect to the main activity of mathematicians at the time. This consisted, in its general outlines, in the positioning of problems and in their solution through a geometric construction.

A first answer can be ventured in light of the peculiar grammatical form in which the impossibility theorem presented in *De quadratura arithmetica* is stated. In asserting that there is no more geometrical quadrature than his own, in fact, Leibniz set a clear-cut limit on the type of exactness with which a solution to the universal quadrature needed to be endowed and, at the same time, advised the mathematician against searching any further for a "more geometrical" solution, which would exhibit the quadrature of a sector through an equation or through a construction by geometrical curves.

Yet, was a solution of this kind, expressed by an infinite series, a solution at all? It certainly was not what one might have expected as a solution to a geometric problem, because it failed to provide a construction obtained by the intersection of curves, a traditional requirement that a solution to a geometric problem would normally have been expected to fulfill.[157]

Aware of this dilemma, Leibniz noted, in the same *Scholium* to Proposition 31 of *De quadratura arithmetica*:

> I don't even promise a quadrature by means of a geometrical construction, but via an arithmetical or analytical expression. Indeed the nature of a series, even infinite, can be understood even if only a few terms are understood, provided the law of formation [*ratio*] of the series is evident. Once this is found, it is useless to continue the series, if the point is for clarifying our understanding instead of performing a mechanical operation. If one asks for a true analytical and general relation which intervenes between the arc and the tangent, one can find in this proposition everything that can be done by Man, as I will prove below [namely, in the last proposition of the treatise]. One can find an equation of a very simple kind which expresses the dimension of the unknown quantity, whereas so far geometers have provided only approximations but not equations for the arc of the circle. I shall be silent on the fact that no one has given rational approximations to any arc or portion of the circle. Therefore, I am now the first by means of whose equation circular arcs and angles can be dealt with by an analytical calculus after the manner of straight lines.[158]

[157]*Schol.* 31, AVII6, 51, p. 600: "At inquies magnitudo quaesita sic non potest exhiberi, quoniam in nostra potestate non est progredi in infinitum."

[158]AVII6, 51, p. 600: "At inquies magnitudo quaesita sic non potest exhiberi, quoniam in nostra potestate non est progredi in infinitum. Fateor: neque enim eam constructione quadam geometrica exhibere promitto, sed expressione Arithmetica sive analytica. Seriei enim, licet infinitae, natura intelligi potest, paucis licet terminis tantum intellectis, donec progressionis ratio appareat. Qua semel inventa frustra progredimur, quoties de mente potius illustranda, quam de operatione quadam mechanica perficienda agitur. Itaque si quis veram relationem analyticam generalem quaerit quae

Thus, according to Leibniz, it is sufficient to know the law of formation of an infinite series for the whole series to be exactly known. One could certainly understand a series in geometrical terms, namely, as a rule for performing approximate constructions. However, Leibniz also made it clear that these constructions need not be executed in order to have a better understanding of the series itself, although they can also serve for practical purposes. It was in this sense that I read Leibniz's remark above, concerning the uselessness of continuing (calculating) the terms of the series in order to elucidate our understanding of the series itself ("…*de mente illustranda*"). On the other hand, by calculating successive terms of the series (or performing the related constructions), one enters the realm of mechanical operations, useful for the practical goal of performing trigonometrical calculations without tables, and with an error as small as we please.[159]

With this consideration in mind, we might relate back to the impossibility of the universal quadrature of the central conic sections the following remarks, made by Leibniz to Conring while discussing the particular quadrature of the whole circle:

> Perhaps my quadrature shall be published one day in France, where I left my proofs. It is not the one desired by the vulgar mathematicians, but the one they should desire. Indeed it is impossible to express by one number the ratio between the circle and the square, but an infinite series of numbers continued to the infinite is necessary, and I think a simpler series than mine cannot be given.[160]

The impossibility of finding a perfect quadrature, that is to say, the one actually desired by the "vulgar" practitioner (although even more refined mathematicians, like Huygens, Leibniz's mentor and correspondent, believed in the possibility of the perfect quadrature of

---

inter arcum et tangentem intercedit, is quidem in hac propositione habet, quicquid ab homine fieri potest ut infra demonstrabo. Habet enim aequationem simplicissimi generis quae incognitae quantitatis magnitudinem exprimit cum hactenus apud geometras appropinquationes tantum, non vero aequationes pro arcu circuli demonstratae extent. Ut taceam ne appropinquationes rationales cuilibet arcui aut portioni circulari communes a quoquam fuisse datas. Quare nunc primum hujus aequationis ope arcus circulares, et anguli instar linearum rectarum analytico calculo tractari possunt."

[159]Cf. the same *Scholium*, AVII6, 51 p. 600: "et si quando contemplationem ad praxin referre licebit, operationes trigonometricae, ingenti geometriae miraculo sine tabulis perfici poterunt, errore quantumlibet parvo."

[160]The letter was written on March 19, 1678. Cf. AII, 1, p. 606: "Tetragonismus meus edetur fortasse aliquando in Gallia, ubi demonstrationes reliqui. Non est qualem desiderant Mathematici vulgo, sed qualem desiderare debent; nam rationem inter Circulum et Quadratum uno numero explicare impossibile est, opus est ergo serie numerorum in infinitum producta, nec puto simpliciorem dari posse quam mea est." It should be pointed out that no conclusions can be drawn, on the grounds of *De quadratura arithmetica*, concerning the quadrature of the whole circle.

the circle), establishes that the arithmetical quadrature is the solution that mathematicians *should* desire, not the one they do desire.[161]

Let us recall that, according to the procedure established in Descartes' *Géométrie*, an equation was by no means the solution to a geometric problem, but was rather the last step in its analysis, which needed to be supplemented by a construction through the intersection of geometrical curves. By contrast, in the case of the quadrature of the sectors of a central conic, an equation—i.e., an infinite series—constitutes in itself the solution to the original geometric problem. It certainly has a geometric, constructive meaning, but this is downplayed as a mere aid to practice.

This is perhaps one of the major conclusions that Leibniz derived from the impossibility of providing a "perfect" quadrature of the circle and the other central conic sectors. In some cases, it may not be possible to exhibit the sought-for object by means of a finite stepwise construction. But even if a traditional solution is impossible in such cases, it does not follow that an exact solution is impossible, so long as we rethink the concepts of exactness and geometricity. Specifically, Leibniz advocated for the legitimate use of infinite series both as instruments in problem-solving and, in some cases, as genuine solutions to problems themselves, like that of the universal quadrature of the circle.

[161] In this sense, Leibniz anticipates a viewpoint on the role of impossibility statements that would be emphasised in the Eighteenth Century, with Montucla and Condorcet (see Lützen 2014, pp. 244–245).

Conclusion 4

4.1 The Role and Goals of Impossibility Results in Early Modern Geometry

In the previous chapters, I have shown that the question about the solvability of the problem of the indefinite and definite quadratures of the circle, the ellipse, and the hyperbola was a source of studies and debates among geometers from the second half of the Seventeenth Century. More particularly, I have chosen to study one episode within precise temporal bounds (1667–1676). The example I chose to analyze revolves around the claim made by James Gregory, in his *Vera circuli et hyperbolae quadratura*, regarding the impossibility of solving the quadrature of the circle in an exact way, and the criticism to which this claim was later exposed.

The importance of this episode in the history of mathematics has been recognized by historians for a long time. It is an early example of a debate centered on the solvability of a problem, rather than a particular attempt to find a solution. It is no wonder that it rang a bell among Nineteenth and early Twentieth Century mathematicians and historians, who were accostumed to analogous results from the mathematical practice of their century.[1]

However, while more traditional historical studies have mainly stressed the aspect of novelty inherent in these early impossibility results and focused on the question of the validity of Gregory's theorem with respect to our own mathematical knowledge, my approach has been one that considers that and similar theorems on their own terms. In short, understanding the above results on their own terms amounts to an inquiry that focuses on how the authors proved their own results, how they understood those results, and the objections raised by other mathematicians from the era in question. In this book,

[1]For examples of this episode's treatment in the early Twentieth Century, see Zeuthen (1903), Dehn and Hellinger (1943) and Dehn and Hellinger (1939).

D. Crippa, *The Impossibility of Squaring the Circle in the 17th Century*,
Frontiers in the History of Science, https://doi.org/10.1007/978-3-030-01638-8_4

I have adopted an approach that uses extant sources to investigate these topics. A first general conclusion emerging from my study is that the problems of the indefinite and that of the definite quadratures of the central conic sections were conceived as distinct problems only as a result of a process lasting several years, to which the controversy between Gregory and Huygens largely contributed. A second general conclusion is that the belief in the possibility of expressing π as a fraction or as an algebraic number, which would be viewed today as a misconception or a consequence of poor knowledge of mathematics, appeared as a reasonable belief in the middle of the Seventeenth Century. It was embraced unhesitatingly by Huygens, for example, who certainly cannot be ranked among amateur mathematicians. Probably through Huygens' influence, Leibniz considered the same belief to be plausible too, at least in a first stage of his career, as I have discussed in the previous chapter. In this regard, I stress that, even though an impressive scholarly effort has made *De quadratura arithmetica* a well-known text, there are still features that have yet to be remarked upon. For instance, as I have shown in my study of the drafts leading up to the final manuscripts, one of the goals of Leibniz's project until the end of 1674 was the search for a solution to the quadrature of the circle by expressing π as a ratio between two integers or as a surd number.

The theoretical framework by reference to which the authors studied in the previous chapters considered the quadrature of the central conic sections to be a solvable or an unsolvable problem was represented by Descartes' geometry. As briefly recalled in the introduction of this book, Descartes' method of problem-solving essentially consisted of two components: a reduction of a problem to a finite algebraic equation; and a geometric construction of the equation by algebraic curves. Descartes also declared that no problem involving the comparison between segments and arcs, like the quadrature of the circle, was reducible to this canon. This conclusion does not entail that no solutions to the circle-squaring problem exist absolutely. For example, it was still possible to find approximate numerical solutions or constructions by mechanical curves, as the ancients did.

From our vantage point, we consider Descartes' belief to be correct in the case of the rectification of the circumference of the circle, since the problem is algebraically unsolvable. However, the kind of knowledge required to prove this result was unavailable to mathematicians from the second half of the Seventeenth Century. On the contrary, Descartes' belief started to waver as soon as it was shown that a class of algebraic curves could be rectified algebraically, among which was the semicubical parabola rectified algebraically by Van Heuraet. To obtain the rectification of the semicubical parabola, Van Heuraet violated the requirement of Cartesian analysis to study, in a given problem, only those relations obtaining segments of finite length. On the contrary, he employed infinitesimalistic techniques, by using rectangles or line-segments of infinitesimal breadth or length, as part of his proof. In spite of such violation of one of the Cartesian methodological tenets, the rectification of the semicubical parabola was printed, together with other essays containing research on Descartes' *Géométrie*, in the second Latin edition of that text. If Van Heuraet's procedure was not seen as a shocking deviation from the Cartesian canon, this was perhaps for the following reason: the appeal to infinitary and infinitesimal objects was not viewed as problematic, provided their use led to finite solutions expressed in an algebraic form. Thus, in Van Heuraet's proof, infinitesimals were useful tools for the analysis of the problem, but did not play any role in the synthetic part, as, in order to construct a straight line equal to the given curve, one would only need algebraic curves.

A similar example in which infinitary objects have only the role of heuristic instruments is the quadrature of the parabola. In this case, an infinite progression, the geometric series $\sum_{n=0}^{n=\infty} \frac{1}{4^n}$, can be used during the process of analysis to discover that the ratio between the parabolic segment and the corresponding triangle is equal to $\frac{4}{3}$. However, once this ratio is known, the parabola can be squared using ruler and compass on the basis of this outcome of analysis. The infinite progression $\sum_{n=0}^{n=\infty} \frac{1}{4^n}$ is confined to the heuristic part of the problem-solving strategy, that is, to the analysis, and, since the sum of this series is a finite quantity, in the synthesis, a "classical" constructivist approach can be taken in order to construct the desired solution.[2]

These examples did not undermine the constructivist aspect of Cartesian geometry, which still survived in the construction of the problem. Only curves that are considered legitimate are used in the construction stage, whereas infinitary objects are duly eliminated at this stage. In this way, one would be justified in using infinite series or infinitesimal magnitudes as tools of reasoning prior to the construction stage of the problem, and the exact and finite solution of certain quadrature or rectification problems would still be acceptable within Cartesian geometry. It is possible that geometers like Huygens were convinced that the circle-squaring problem could be solved in a similar way employing an infinite analysis, which would result in a finite geometrical construction in the synthesis. In other words, they could have been convinced of the completeness of the Cartesian method, in the sense that any problem that we would consider geometrical would be susceptible to receiving an exact solution, understanding "solution" in the sense established by Descartes. After all, if the circle and the hyperbola are both geometric curves, and many of their properties (such as their equations, tangents, ...) can be found using Descartes' geometric algebra, why not also the areas under such curves?

These considerations also explain very well the rationale behind Gregory's inquiry, presented in the preface of the *VCHQ*. He there questions whether Cartesian geometry could tackle and solve quadrature problems in all of their generality, hence the quadrature of the circle too. In spite of repeated efforts, the case of the quadratures of the circle and the other conic sections with a centre stood out as problematic, since neither the known infinitesimal methods nor the classical polygonal approximations led to constructions by

[2]Archimedes was ill at ease with computations involving infinite series, thus he used the more rigorous method of exhaustion. Several centuries later, Viète studied the same theorem, and hinted at the possibility of using the geometric progression as a shorter method of proof (Viète 1593, p. 29). Like Van Heuraet's proof, the non-finite portion of Viète's solution only occurs during the heuristic step of analysis, and the fact that the infinite sum reaches a finite value allows the synthetic portion of the solution to be carried out by a "classical" constructivist approach.

geometric curves that were acceptable by Descartes' standards. Gregory explicitly posed the question, and came out with a negative answer in the form of an impossibility result, which we have discussed at length in the first chapter of the book.

Because of the impossibility of expressing the area of any sector as an analytical composition of the in- and circumscribed polygons, i.e., of the radius and the chord, a new infinite operation had to be admitted besides the five arithmetical ones. Gregory's sixth operation, namely, the construction of convergent series, formalised the Archimedean construction of two successions of polygons of increasingly many sides to a given sector.

In the introduction to *Geometriae Pars Universalis* (1668), a work that I have not discussed in this book, Gregory concluded that Cartesian analysis as a method was simply unable to deal with the "measurement of curvilinear quantities."[3] That is to say, the Cartesian method cannot solve all problems of area and rectification of curvilinear figures, echoing the considerations raised in the VCHQ. In these cases, one would need to use non-Cartesian objects, like infinite series, as a "final resource" (*ultimum nostrae methodi refugium*).[4] The doctrine of infinite convergent series, Gregory adds in the same preface, is general, because it always exhibits a possible solution of a problem, starting from the properties of a given figure. Thus, it makes the method of Cartesian geometry complete, or, in any case, constitutes a decisive step towards the discovery of a method capable of achieving this goal.

Along similar lines, a few years later, Leibniz criticized the "short-sightedness" of Cartesian geometers. He invoked the need to search for a method that would represent for the study of quadratures and rectifications what Viète's and Descartes' symbolic algebra had represented for "Apollonian" geometry, that is to say, the study of loci of algebraic curves.[5] Such a judgment was strongly supported by Leibniz's impossibility theorems, which close the final draft of *De quadratura arithmetica*. These theorems prove that the problem of the indefinite quadrature of the circle and the hyperbola do not fall within the bounds of Cartesian geometry.

However, as we have seen in the previous chapter, Leibniz seriously considered the idea that his series for the area of the circle (namely, for $\frac{\pi}{4}$) might converge to a known rational or irrational number. This belief does not contradict Leibniz's impossibility results about the indefinite quadrature of the circle. This is because Leibniz accepted Huygens'

[3] *GPU*, p. 3.

[4] "And if the geometer, after the proper application of this method according to the properties of the figure, cannot find any solution, he must resort to convergent series, whose limit is the unknown figure or another one, in a given ratio with this one." See Malet (1996, p. 226), for a further discussion.

[5] Cf., among several manuscripts, VII6, p. 504: "nam cum linearum curvarum, aut spatiorum ipsis conclusorum magnitudo quaeritur (...) neque aequationes neque curvae Cartesianae nos expedire possunt; opusque est novi plane generis aequationibus, constructionibus curvisque novis; denique et calculo novo, nondum a quoquam tradito, cujus si nihil aliud saltem specimina quaedam, mira satis, jam nunc dare possem."

objections against Gregory and, briefly speaking, explicitly recognised a fact that we could state as follows, albeit slightly anachronistically: transcendental functions such as trigonometric or logarithmic ones may take algebraic values for some of their arguments. Thus, for Leibniz and for Huygens, the problems of the indefinite and definite quadratures of the circle were considered as distinct problems, in the sense that the impossibility of the former did not entail the impossibility of the latter.

Nowadays, we know that the definite quadrature of the circle is impossible, because π is not a rational number (the first proof is credited to Lambert), or an algebraic one (this result was proven by Hermite and Lindemann), hence the area of the whole circle is transcendental with respect to the radius. Without the necessary mathematical instruments to prove either result, Leibniz could, at most, advance conjectures about the unsolvability of the definite quadrature of the circle and the nature of π. Yet, even if it remained undecided as to whether the definite quadrature of the circle could be solved within Cartesian geometry, this is not the end of its story. In fact, Leibniz claimed, on several occasions, to have found an exact solution to the circle-squaring problem. Such a solution was represented by his arithmetical quadrature, that is to say, the infinite series:

$$\frac{1}{1} \pm \frac{1}{3} \pm \frac{1}{5} \pm \frac{1}{7} \pm \ldots$$

As I have argued in the previous chapter, Leibniz's conclusion is justified on the basis of an idea of exactness different from the Cartesian one. The latter was based on finite algebra and constructibility through geometric (algebraic) curves. To stress the contrast, let us note that the arithmetical quadrature of the circle possesses the following two features in contrast with Cartesian geometry: (1) it is given by an infinite series of rational numbers; (2) it is not—or rather does not need to be—supplemented by a construction through geometric curves, although the series might be interpreted as a rule for finding an approximate geometrical solution to the quadrature of the circle.

The reason why Leibniz considered that an infinitary object like the series for $\frac{\pi}{4}$ represented an exact solution is well-explained in several drafts of *De quadratura arithmetica*: in brief, even if the sum of the series expressing the area of the circle is obtained by an infinite iteration of arithmetical operations, the ratio of the series can be perfectly known by the intellect, as it is expressible in finite terms, namely, as $\sum_{k=0}^{\infty} \frac{-1^k}{2k+1}$. This rule that governs the progression of the series guarantees the exact knowledge of the sought-for object; although it represents a different kind of knowledge than the one obtained by a finite algebraic expression, it is nevertheless the best attainable type of knowledge with respect to the problem at hand.[6]

We find analogous considerations in *De vera proportioni circuli ad quadratum circumscriptum in numeris rationalibus expressa* (1682), an article written several years after completion of *De quadratura arithmetica*, in which Leibniz summarised much of the

[6]See also Debuiche (2013).

work on the problem of the quadrature of the circle done between 1673 and 1676. In particular, Leibniz compares, in this article, his own discovery with the previous history of the quadrature of the circle and with the attempts to calculate π:

> Archimedes, by inscribing and circumscribing to the circle, because the latter is greater than inscribed polygons and lesser than circumscribed ones, shows a way of presenting the limits, between which the circle must fall, or of showing the approximations: clearly the ratio of the circumference to the diameter is greater than 3 to 1 or than 21 to 7, and less than 22 to 7. Others have pursued this method, Ptolemy, Vieta, Metius, but especially Ludolph van Ceulen, who showed the circumference to the diameter to be as 3.14159265358979323846 etc. to 1.00000000000000000000. Truly approximations of this kind, even if they have practical uses in geometry, yet show nothing, which may satisfy the mind in great need of the truth, unless a progression of such numbers being considered to infinity may be found.[7]

From the viewpoint of practical geometry, the Archimedean method for squeezing the circumference between two sequences of polygons and truncating the double sequence in order to find finite approximations is even more valuable than Leibniz's arithmetical series, because the latter converges slowly. However, the aim of Leibniz's research was not practical, but rather moved from what we could consider a foundational concern. Specifically, Leibniz sought an exact solution that could be grasped with the same simplicity, clearness and distinctness of an algebraic solution:

> Even if it is not possible to express the sum of this series by a single number, and the series is indefinitely continued, it can all be understood well enough, because it yet agrees with a single law of progressing. For if indeed the circle is not commensurable with the square, it cannot be expressed by a single number, but it must be shown to be of necessity by a series in terms of rational numbers. Similar cases are the diagonal of the square, or a segment cut in mean and extreme ratio, that some call divine, and many other quantities, which are irrational.[8]

---

[7]Leibniz (2011, pp. 8–9): "Archimedes quidem Polygona Circulo inscribens & circumscribens, quoniam major est inscriptis, & minor circumscriptis, modum ostendit, exhibendi limites intra quos circul[u]s cadat, sive exhibendi appropinquationes: esse scilicet rationem circumferentiae ad diametrum, majorem quam 3 ad 1, seu quam 21 ad 7, & minorem quam 22 ad 7. Hanc Methodum alii sunt prosecuti, Ptolemaeus, Vieta, Metius, sed maxime Ludolphus Coloniensis, qui ostendit esse circumferentiam ad diametrum, ut 3.14159265358979323846*&c.* ad 1, 00000000000000000000. Verum hujusmodi Appropinquationes, etsi in Geometria practica utiles, nihil tamen exhibent, quod menti satisfaciat, avidae veritatis, nisi progressio talium numerorum in infinitum continuandorum reperiatur."

[8]Leibniz (2011, p. 10): "Et licet uno numero summa ejus seriei exprimi non possit, et series in infinitum producatur, quoniam tamen una lege progressionis constat, tota satis mente percipitur. Nam siquidem circulus non est quadrato commensurabilis, non potest uno numero exprimi, sed in rationalibus necessario per seriem exhiberi debet; quemadmodum & Diagonalis quadrati, & sectio extrema & media ratione facta, quam aliqui divinam vocant, aliaeque multae quantitates, quae sunt irrationales."

The measurement of the circumference can thus be known with exactness ("it can all be understood well enough") through Leibniz's series because its ratio is known, namely:

$$\sum_{k=0}^{\infty} \frac{-1^k}{2k+1} = \frac{\pi}{4}.$$

The analogy made by Leibniz between the area of the circle and the length of the diagonal of a square further illustrates his ideal of geometrical exactness. Being irrational, $\sqrt{2}$ (or, interpreted geometrically, the measure of the diagonal of a square with sides of unit length) could be represented through its rational approximations calculated through a suitable numerical algorithm, such as the Euclidean algorithm of continual subtraction applied to incommensurable magnitudes. On the other hand, $\sqrt{2}$ can also be expressed through an infinite series by applying the binomial theorem, as Leibniz was certainly aware.[9] In this way, we obtain

$$\sqrt{2} = (1+1)^{\frac{1}{2}} = \sum_{k=0}^{\infty} \binom{\frac{1}{2}}{k} (1)^k.$$

This is, in turn, equal to the infinite series:

$$= 1 + \frac{1}{2} + \frac{1}{8} + \frac{1}{16} + \ldots$$

From the perspective of exactness, the crucial distinction between the process of determining $\sqrt{2}$ by an iterative algorithm and that of expressing the same number as an infinite sum-series does not depend on which method provides a better numerical approximation. On the contrary, the former procedure, independently of how fast the desired number approaches, is always bound to provide the knowledge of a single $(n+1)$-th approximation of $\sqrt{2}$, given the previous step n.[10]

The binomial expansion expresses the irrational quantity $\sqrt{2}$ through a formula that contains the ratio of the series, and therefore can represent all of the infinite operations of addition and subtraction that yield $\sqrt{2}$.

[9]Cf. AII2, p. 252.

[10]The fact that this procedure could not lead to true knowledge was stressed by Stevin: "Encore qu'il nous fust possible, de soubstraire par action, plusieurs cent mille fois la moindre grandeur de la majeure, & le continuer plusieurs milliers d'années, toutefois (estant les deux nombres proposez incommensurables) l'on travailleiroit eternellement, demeurant toujours ignorant, de ce qui encore à la fin en pourroit encore avenir. Cette manière donc de cognition n'est pas légitime, ainsi position de l'impossible, afin d'ainsi aucunement declairer, ce qui consiste veritablement en la Nature ..." (Stevin 1958, Vol. 2, p. 723).

The same difference between exact expression and accurate approximation occurs when we compare the measurement of the area or the circumference of the circle, obtained through approximation algorithms based on the Archimedean method, with the arithmetical series obtained by Leibniz. As Leibniz argues in *De vera proportioni circuli*, only through the latter are we to understand a "true" (i.e., exact) solution to the quadrature of the circle.

The proof that the circle-squaring problem may not be solved by finite methods justified Leibniz's search for new criteria of exactness suitable for transcendental problems. A rather opposite attitude was taken by Gregory, who simply dismissed any attempt to define exactness as depending upon particular methods. Perhaps a tacit lesson that he gained from impossibility results was that there is no ultimate, normative canon of exactness, as, in principle, one cannot foresee the emergence of new impossibility results that may overthrow a given canon. As a result, Gregory was to adopt the liberal solution of identifying the most exact method with the simplest one, depending on the problem one has to solve, although "simplicity" has no fixed meaning. We find echoes of Gregory's attitude towards exactness and simplicity in Wallis, for instance, who had no qualms in rejecting the Cartesian distinction between geometrical and mechanical curves as a pseudo-problem:

> Whether he will call Geometrical, all which may be imagined to be thus constructed; will be at his choise: I list not to dispute of names: Since it is at the liberty of a Mathematician, to determine by a Definition, at what latitude or narrowness he wil have such a word (with him) to be understood. And so he may call those lines which I mentioned, Geometrical or not-Geometrical, as he pleaseth.[11]

Thus, the question of what is exact, and therefore geometrical, and what is not became, for Wallis, as well as for Gregory, a matter of personal taste or of convention, which lay outside mathematics.[12]

4.2 Two Early Modern Conceptualisations of Impossibility Results

The impossibility of squaring the central conic sections algebraically was used by early modern mathematicians in order to settle certain issues that we could term "meta-mathematical," and which regarded the limits of applicability of Descartes' method. Particularly in the above section and, more extensively, in the previous chapters, I have pointed out two methodological consequences that the impossibility of the algebraic quadrature of the central conic sections had for the problem of establishing the bounds of Cartesian geometry: (a) impossibility results accounted for the incompleteness of Descartes' method of problem-solving, i.e., the latter could not provide a solution for all

[11] Wallis (2014, p. 426).

[12] A similar view might have influenced Newton, as shown in Blåsjö (2017, pp. 56–57).

problems of geometry, contrary to Descartes' expectations; (b) as a result of the non-algebraic quadrability of the central conic sections, the traditional requirement according to which a problem must be solved by a geometric construction, performed in finitely many steps, began to lose its compelling methodological force. In its place, infinite series were admitted as legitimate solutions for problems irreducible to finite algebraic equations; as we have seen in the previous section, Leibniz explicitly argued for new criteria of exactness, alongside the exactness norms of Cartesian geometry.

To conclude this study on impossibility results in the early modern period, I shall survey two methodological discussions on impossibility results in mathematics. The general purpose of this survey is to show that impossibility results were not considered as isolated cases during the second half of the Seventeenth Century, but were studied as part of a more general mathematical phenomenon.

As already mentioned in Chap. 1 of the present work, both Gregory and Wallis entered methodological discussions on the impossibility of squaring the central conic sections. For both of them, the impossibility of squaring the circle by known, legitimate means caused the search for new kinds of mathematical object that could express the solution. From Gregory's viewpoint, such a solution could be given by a new kind of operation, which he called the "sixth operation." Wallis, on the other hand, relied on the possibility of creating a new kind of numerical notation in order to express the quadrature of the circle. In his *Treatise on algebra* (1685), Wallis resumed and expanded the discussion about impossibility in mathematics given in the *Arithmetica infinitorum* by providing a systematic account of the role impossibility played in arithmetic, and therefore in the whole of mathematics of which, according to him, arithmetic represented the ultimate foundations.[13]

Wallis divided arithmetical operations into two classes: synthetical or genetical and analytical or resolutive.[14] Addition is a synthetical operation: we start from two given numbers a and b and obtain a third number c by adding a to b. Similarly, multiplication and exponentiation are synthetical operations as well: the former can be thought of as a repeated addition, the latter as a repeated multiplication. Whereas addition is the operation involved in the problem of finding a quantity c given two quantities a and b, subtraction is defined as the operation involved in the problem of finding a quantity b, given a quantity a and the sum $c = a + b$. In this sense, it is the reverse of addition, because it is applied when the result of addition is known, and the givens have to be found. Probably for this reason, Wallis called the operation of subtracting one number from another as an "analytical" or "resolutive" operation. A similar reasoning applies to division (with respect to multiplication) and to root extraction (with respect to exponentiation). Wallis also managed to include, in the same general scheme, the extraction of complex roots

[13] See also Jesseph (1999, pp. 37 and 38).

[14] Wallis (1685, p. 316).

that arise in the activity of solving algebraic equations, which, for Wallis, belonged to the analytical part of mathematics.

So much for the usual arithmetical operations. To these, according to Wallis, another pair of operations may be added: the synthetic operation consisting in the composition of a geometric progression, such as: $1, 4, 16, 64 \ldots$, and the corresponding analytical operation of interpolating a geometric progression. This is an operation on progressions amply used by Wallis in the *Arithmetica infinitorum*: given a progression such as: $1, 4, 16, 64 \ldots$, by interpolation, we insert new terms between the given ones, obtaining, for example, the new progression: $1, \mathbf{2}, 4, \mathbf{8}, 16, \mathbf{32}, 64 \ldots$ The operation can be repeated, so as to obtain still another progression: $1, \boldsymbol{\sqrt{2}}, 2, \boldsymbol{2\sqrt{2}}, 4 \ldots$[15]

According to Wallis, impossibilities naturally arise in the resolutive, or analytical, part of arithmetic:

> Addition is Genetical, (or synthetical), and to any positive number, any positive Number may be Added, without coming to any Impossibility. But Subduction is Analytical, or Resolutive: and here the case is sometimes possible; as if a Lesser be to be subducted from a Greater; $(3 - 2 = 1)$ But sometimes Impossible; as if a Greater be to be taken from a Lesser; $(2 - 3)$ in which case we are provided of a Notation, to express that Impossibility (and the measure of that Impossibility) by a negative Quantity $(-1 = 2 - 3,)$ imparting somewhat less than Nothing.[16]

The impossibility of performing the operation of subtraction arises only in the background of the Euclidean notion of an integer number as only constituting the positive integers. In a similar way, dividing mutually prime integers results in another impossibility, because no natural number can result from such an operation, and similar cases of impossibility can emerge in the case of root extractions. Wallis understood the negative "number" -1 as an "expedient," namely, a sign that denotes the very impossible solution of the equation: $2 - 3 = x$; similarly, the fraction $\frac{3}{4}$ would denote, for him, the result of the impossible division of 3 by 4; and the surd number $\sqrt{8}$ would be a sign denoting the impossible result of the operation of taking the square root of the number 8. Imaginary quantities were also conceived as arising from the analytical part of arithmetic. In fact, a square root of a negative quantity, such as $\sqrt{-1}$, may be thought of as the sign denoting the impossible operation of finding a mean proportional between the numbers 1 and -1, that is to say: $1 : x = x : -1$.[17] Using the same model, Wallis also managed to account for the impossibility of solving the quadrature of the circle. In the *Arithmetica infinitorum*, the problem of finding the ratio between the quarter of a circle and the square built on the radius, or the diameter, is reduced to the arithmetical problem of interpolating the progression: $1, \frac{3}{2}, \frac{15}{8}, \frac{105}{48} \ldots$. Since the operation of "squaring" the circle is not, for Wallis,

---

[15]The terms in bold represent the terms obtained by successive interpolations.

[16]Wallis (1685, p. 316).

[17]Cf. Wallis to Leibniz, 6/01/1698 or 1699 (LSG, IV, p. 57).

a synthetic operation, but rather corresponds to the analytical operation of interpolating a sequence, it is not surprising that an impossibility result arises in connection with this problem. The impossibility of solving the definite quadrature of the circle should thus be stated in arithmetical terms, as we read in the *Arithmetica infinitorum*: "such proportion [i.e., the proportion between a quarter of the circle and the square built in the radius] is not to be expressed in the commonly received way of notation ..." (Wallis 1685, p. 316).[18]

Even if impossibility results can be definitively proven in arithmetic (such a proof would amount to showing that an analytical operation cannot always be performed), it is algebra that provided the mathematicians with signs for denoting, or "measuring," as Wallis would say, these impossibilities, that is to say, the failure to perform certain operations upon natural numbers. We could even suppose that the correspondence between the invention of new symbols and operations and the appearance of impossibility results constituted a structural law that granted the progress of algebra. As Wallis explained in a letter to Leibniz from the late 1690s, for example:

> If it is supposed that a number n is a fractional, surd or however inexpressible number, new methods of extraction must be contrived, adequate to such cases. Indeed (which I often remark), in all the resolutory operations (like the Subtraction, division, extraction of roots, solution of equations, interpolation, etc.), we always reach what, in a strict sense, cannot be done, but which is however denoted as *quasi-factum* (such are: -1, $\frac{3}{2}$, $\sqrt{2}$ etc). And thus we will continually proceed to other and other steps of *arresias* or inexplicability, to the infinite, since the material to proceed further and further will never lack to those who wish to advance.[19]

In such a way, Wallis came to provide a rational, explicit justification for such entities as surd, negative or imaginary numbers, which had usually been regarded by Renaissance and early modern geometers as "false," "sophistic," "impossible," or "imaginary," although they had nonetheless been employed in practice[20]:

> And therefore what arithmeticians usually do in other work, must also be done here; that is, where some impossibility is arrived at, which indeed must be assumed to be done, but nevertheless cannot actually be done, they consider some method of representing what is assumed to be done, though it may not be done in reality. And this indeed happens in all

[18]Cf. Lützen (2014).

[19]LSG, IV, p. 38: "Quodsi supponatur hoc numerus n, numerus fractus, surdus, vel utcumque *arretos*, comminiscendae sunt novae extractionum methodi casibus hujusmodi congruae. Quippe (quod ego saepe moneo) in omnibus operationibus Resolutoriis (quales sunt Subtractio, Divisio, Extractio radicum, Aequationum solutio, Interpolatio etc.) semper pervenitur ad id quod stricto sensu fieri non potest, sed quod utcumque designetur quasi-factum (ut sunt -1, $\frac{3}{2}$, $\sqrt{2}$ etc). Adeoque continue procedetur ad alios aliosque gradus... seu Inexplicabilitatis, in infinitum, ut nunquam desitura sit materia ultra ultraque procedendi, volentibus id aggredi."

[20]One example that Wallis had in mind was probably that of Cardano and Bombelli: cf. Gavagna (2014, pp. 178–179).

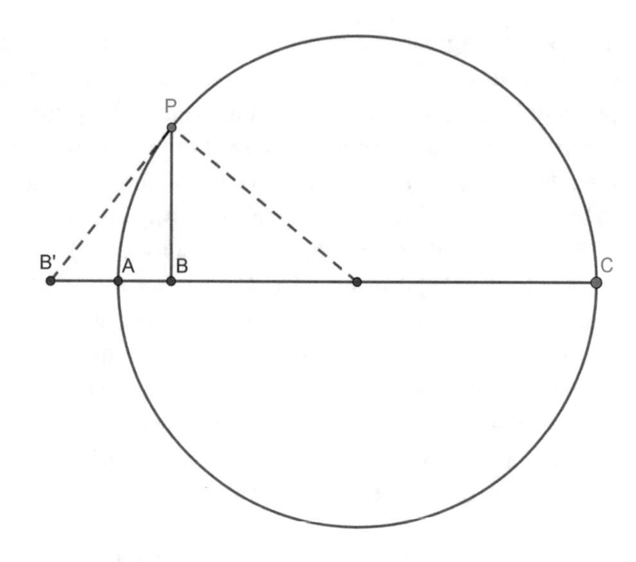

Fig. 4.1 One geometrical interpretation of imaginary quantities given by Wallis (1685, p. 266). Considering the figure, let us assume that $AB = b$, $AB' = -b$ (where B' lies on the tangent to the circle in P), while $BC = c$. Then, we shall have, by elementary geometry, $PB = \sqrt{bc}$, and $PB' = \sqrt{-bc}$

> operations of arithmetic involving resolution, for example, in subtraction: if it is proposed that a larger number must be taken from a smaller, thus 3 from 2 or 2 from 1, since this can not be shown in reality, there are considered negative numbers, by means of which a supposed subtraction of this kind may be expressed, thus $2 - 3$, or $1 - 2$, or -1.[21]

This passage, taken from the *Arithmetica infinitorum*, perfectly resonates with the considerations on impossible operations from the later *Treatise on algebra*.

In that later text, Wallis also made several attempts to ground negative and imaginary quantities by geometrical or physical representations. Thus, negative quantities could be described in a very intuitive way as displacements in a direction opposite to a given, positive direction.[22] On this basis, imaginary quantities can also receive a geometrical interpretation, such as the one described in Fig. 4.1.

Analogous considerations on impossibility in mathematics are also to be found among Leibniz's reflections consigned to his correspondence, especially in the letters exchanged with Wallis, and to his philosophical manuscripts. Particularly interesting and useful as a complement to our discussion are a few philosophical remarks to be found in the *Elementa nova matheseos universalis* (AVI4, 124), a manuscript Leibniz probably wrote at the beginning of the 1680s. In that work, Leibniz distinguished, along the same lines as Wallis, synthetic from analytic operations.[23] He also recognized that there exist analytical mathematical operations that, although they cannot be performed when applied to certain objects, yield a meaningful solution:

[21] Wallis (2004, p. 162).

[22] Wallis (1685, pp. 264–267).

[23] AVI4, 124, p. 518.

> Sometimes the operation which must be performed is actually impossible, either wholly or by lack of time, although it can be exhibited by a construction, at least in our characters, or it is is already exhibited in the nature. Thus, it is impossible to perform a subtraction when nothing remains, and yet this operation is represented in nature, as when one owes more than he possesses. Likewise, it is impossible to actually divide an integer prime number by another. From this division a fraction results, which means that a division must be performed, the thing which is designed by this number being divided in parts more fit to exhibit such division. In the same way incommensurable quantities, that is surd roots, are generated when the extraction cannot occur.[24]

Leibniz's example of subtraction as an actually impossible operation is crystal-clear and requires no comment (as we have seen, Wallis argued along similar lines in his *Algebra*, p. 265, even if he used a geometrical and not a mercantile example). As I understand the above passage, a Euclidean notion of number makes division impossible in arithmetic when the dividend is not a multiple of the divisor. But it can still be possible when thought of as a real process. For instance, in the case of dividing 4 by 5, one decomposes a certain whole into 5 equal parts, of which we subsequently take 4. As for mutually incommensurable quantities, they can nevertheless be represented through geometrical constructions, as was clear since antiquity. It is interesting, with respect to the case studies examined in this book, that Leibniz did include, among incommensurable quantities, not only surd roots, but also non-algebraic quantities like circular arcs and logarithms:

> Incommensurable quantities may also be exhibited in nature as real quantities: these are either algebraic or transcendental. They are algebraic when they are found through the extraction of a root of a certain degree. But they are transcendental when the degree of the equation is either uncertain or inexpressible. And logarithms pertain to transcendental quantities. There are various ways to express a transcendental quantity, either like logarithms, or through certain operations which are supposed to be continued in infinity and that, when they are actually given by their termination after finitely many terms, must approach the desired quantity as much as desired. From this operation approximations are generated.[25]

---

[24]AVI4, 124, p. 520: "Interdum etiam operatio actu ipso facienda vel pro tempore vel omnino est impossibilis, saltem in nostris characteribus etsi construendo exhiberi possit aut a natura jam exhibita sit. Ita impossibile est subtrahi cum nihil adest, et tamen hoc in natura repraesentatur, cum quis plus debet, quam habet in bonis. Item numerum integrum primitivum impossibile est actu dividi per alium; unde fit fractio, quae repraesentat divisionem esse faciendam, re quae isto numero designatur divisa in partes ad eam divisionem exhibendam aptiores. Eodem modo oriuntur quantitates incommensurabiles, seu radices surdae ubi extractio non habet locum." See Leibniz (2018, pp. 98ff.) for a French translation.

[25]AVI, 4, 124, p. 522: "Reales vero licet incommensurabiles quantitates possunt in natura exhiberi: eaeque sunt vel Algebraicae vel Transcendentes. Algebraicae cum inveniuntur extractione radicis certi gradus. At Transcendentes cum gradus aequationis aut est incertus, aut non est enuntiabilis. Et ad Transcendentes pertinent Logarithmi. Et varii sunt modi Quantitatem Transcendentem exprimendi, tum ad instar Logarithmorum, tum certis operationibus quae supponuntur in infinitum continuatae, eae finities actu ipso praestitae quantitati quaesitae quantumlibet accedere debent, unde oriuntur appropinquationes."

As we know from *De quadratura arithmetica*, the relation between a circular arc (respectively, a logarithm) and its corresponding tangent (respectively, number) is not expressible through a finite polynomial equation, but only through an infinite equation—a series, in our terminology. However, just like incommensurable algebraic quantities, arc lengths or logarithms are also exhibited in nature and, for this reason, are real and consistent entities.

Finally, Leibniz claimed that root-extraction of a negative number, which generates imaginary quantities or complex roots of equations, can also lead to meaningful results, even if, unlike negative, fractional and incommensurable numbers or quantities, extracting a root of a negative quantity was not a process somewhat represented in nature:

> For the square root of -1 involves some notion, though it cannot be exhibited, and if anyone wanted to exhibit it by a circle, he would find that the straight line required for this [way of picturing roots] does not intersect the circle. But there is a great difference between problems that are insoluble on account of imaginary roots and those that are insoluble because of their absurdity, as for example, if someone were to look for a number which multiplied by itself is 9 and also added to 5 makes 9. Such a number implies a contradiction, for it must, at the same time, be both 3 and 4, that is, 3 and 4 must be equal, a part equal to the whole. But if anyone were to look for a number such that its square added to nine equals that number times three, he could certainly never show, by admitting such a number that the whole is equal to its part, but nevertheless, he could show that such a number cannot be designated.[26]

Leibniz makes, in the above passage, an interesting distinction between problems "insoluble on account of their absurdity" (*ob absurditatem*) and problems "insoluble on account of imaginary roots" ("*ob radices imaginarias*"). The former case is represented by the example of finding an integer number x that satisfies both of the following conditions: $x^2 = 9$ and $x + 5 = 9$.[27] Obviously, the system

$$x^2 = 9$$
$$x + 5 = 9$$

[26]AVI, 4, 271, p. 1448: "Nam $\sqrt{-1}$ aliquam notionem involvit, licet ea non possit exhiberi, et si quis eam circulo exhibere volet, inveniet circulum illum a recta quae ad hoc requiritur non attingi. Multum tamen interest inter quaestiones quae insolubiles sunt ob radices imaginarias, et quae [in]solubiles sunt ob absurditatem. Ut si quis quaerat numerum, qui in se ductus faciat 9: et ad 5 additus faciat etiam 9; talis numerus implicat contradictionem, debet enim simul esse 3 et 4, seu 3 et 4 debent esse aequales pars toti. Verum si quis numerum quaerat talem cujus quadratum additum ad 9, faciat idem quod numeri triplum. Is quidem nunquam ostendet totum esse majus sua parte, tali numero admisso, sed tamen et illud ostendet, talem numerum non posse designari." (English translation in Leibniz 1984, p. 21).

[27]Cf. also AVI4, 124, p. 521; cf. AVI4, 271, p. 1448.

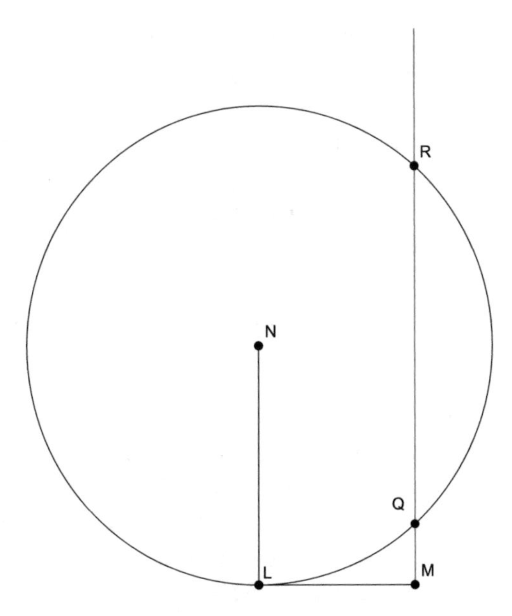

Fig. 4.2 The power of imaginary quantities to represent impossible geometrical configurations was well-known to early-modern geometers. This is the geometrical construction of the equation $z^2 = az - b^2$, when $\frac{1}{4}a^2 < b^2$, according to Descartes. In this case, the segment $NL = \frac{1}{2}a$ will be smaller than the segment $LM = b$, so the circle and the segment MR fail to intersect, and the equation has no geometrical solution, following Descartes' protocol. Analytically, this situation corresponds to the formula: $z = \frac{1}{2}a \pm \sqrt{\frac{1}{4}a^2 - b^2}$. A root of a negative quantity thus appears in the analytical expression

cannot have any solution for x, as the unknown should be equal to 3 and 4, which is absurd, since, 3 being a part of 4, admitting their equality would violate the axiom that the whole is greater than the part.

Leibniz demarcates this absurd case from the case of problems leading to imaginary solutions or roots. The latter may be exemplified by certain impossible problems in geometry, such as when we are searching for the intersection between a straight line and a circle, although these two lines fail to intersect in the geometric plane (see also Fig. 4.2 below).

In such a case, a solution that is not representable geometrically in the plane can still be represented symbolically by imaginary quantities (it is possible that, when he wrote this, Leibniz was unaware of Wallis's attempt to represent such solutions on a plane). It may correspond, for instance, to a quadratic equation in which the discriminant is negative, as in the equation $x^2 + 9 = 3x$ mentioned in the passage above. The roots cannot designate any real quantity there, yet they do not violate Euclid's overarching axiom about the whole and the part, and therefore they should not be considered absurdities.

For Leibniz, then, the result of extracting a square root (or another even root) of a negative quantity engenders "accidental" or "apparent" impossibilities, that is to say, numbers or quantities that are actually the result of analytical operations, when the latter are illegitimately employed beyond their usual area of application.[28] Apparent impossibility should not stop us from treating "impossible" quantities or numbers as if they were real quantities, even when they cannot be represented. There is also a pragmatic aspect in Leibniz's acceptance of apparent impossibilities, since imaginary quantities obey arithmetical rules and imaginary roots can be counted together with real ones, in order to secure the generality of the fundamental theorem of algebra, at least in the version known to early modern mathematicians.[29]

From the previous survey on Wallis's and Leibniz's views, the following conclusion can be drawn: both mathematicians admitted the existence of mathematical impossibilities, which should be distinguished from mere logical impossibility. While the latter are notions, results or statements that are logically inconsistent, and therefore cannot have any content, the former have meaningful content. Such meaningful content is granted either by the activity of mathematicians or by a geometrical or physical interpretation. Particularly, mathematically impossible quantities denote the results of analytical operations whenever certain operators are extended beyond their classical constraints of application. As we have seen, such cases occur when we subtract a quantity from a quantity less than it, or extract a root of a number that is not a perfect square or the square root of a negative number, or, again we measure a curvilinear magnitude, and so on. Such symbols as -1, $\sqrt{2}$, $\sqrt{-1}$, etc. thus have a sound meaning. According to Wallis, they denote the very impossibility of solving given problems. Leibniz had perhaps a less clear-cut position on their reference; however, he was convinced that they were merely "accidentally" impossible. There were three reasons for this: first, they could be interpreted as denoting real processes; second, they were subject to sound algorithms (a theme also present in Wallis); and finally, they did not violate the fundamental axioms of the part-whole relation in any perceivable way.

The distinction between logical impossibilities and mathematical impossibilities was not mere speculation, as it was to be discussed among later mathematicians too, most notably Gauss.[30] It seems to me that the origins of the idea that mathematics can "host"

[28]For a further discussion, see Cortese (2016).

[29]AVI4, 124, p. 520: "Et quaedam extractiones tales sunt, ut radices illae surdae nec in natura rerum extent, tunc dicuntur imaginariae ... utile tamen est eam perinde considerari ad realem, quia radices imaginariae computandae cum realibus; ut integer numerus radicum alicujus aequationis habeatur; aliaque multa de ea utilia possint determinari." A similar argument can be found in Girard's *Invention nouvelle en l'Algebre*: "Donc il se faut resouvenir d'observer tousjours cela: on pourroit dire à quoy sert ces solutions qui sont impossibles, je respond pour trois choses, pour la certitude de la reigle generale, & qu'il ny a point d'autre solutions, & pour son utilité"(Girard 1629, p. 47; also quoted in Gavagna 2014, p. 186).

[30]Let us recall, for instance, Gauss's polemical remarks against the common habit of calling certain numbers or roots "impossible" when they are not (logically) impossible at all. According to Gauss, this is bad terminology because complex numbers are not logically impossible, unlike, say, "rectilinear equilateral right-angle triangles": "If someone would say a rectilinear equilateral right triangle is impossible, there will be nobody to disagree. But, if he intended to consider such an

contentful impossibility statements are to be traced back precisely to the development of algebra and geometry during the Sixteenth and Seventeenth Centuries, including the debates around the impossibility of squaring the circle and the central conic sections.

4.3 Concluding Remarks

From our overall discussion, we can extrapolate the following general conclusion: the question of impossibility became central in the work of many outstanding early modern mathematicians. This interest in impossibility may suggest parallels with later historical periods, when impossibility results also received special attention and were sometimes discussed using a similar terminology and examples. It is sufficient to think of the proofs that the three classical problems of antiquity are unsolvable by ruler and compass, of the proof of the impossibility of solving the quintic, or also Joseph Liouville's investigations of the integrability of functions and differential equations in finite form. All of these results were given during the Nineteenth Century.[31] In all of these cases, let us recall, abstract algebra played a central role in establishing a priori, that is to say, before even searching for a solution, the formal conditions of solvability and unsolvability of problems.[32]

However, as I have argued elsewhere, such parallels should be taken *cum grano salis*.[33] Certainly, one of the greatest episodes of early-modern mathematics was the surge of symbolic algebra and its applications to geometry. But the nexus between impossibility results and early-modern algebra appears at two levels that are specific to early-modern mathematics. First, on a purely operational level, algebra offered mathematicians a useful means in order to devise geometrical impossibility proofs. For instance, both Gregory's and Leibniz's arguments for the impossibility of squaring the central conic sections rely on algebraic properties of equations (such as the irreducibility or their degree) and on formal manipulations that do not admit any obvious interpretation in geometry.

On a second, methodological level, it should be pointed out that algebra, at least for many authors, essentially consisted of a method of discovery, namely: "an analytical problem-solving method for arithmetical [and geometrical] problems using an unknown quantity represented by an abstract entity."[34] Algebra was, in other words, an analytical art. As we know from the classical account by Pappus and from the early modern

impossible triangle as a new species of triangle and to apply to it other qualities of triangles, would anyone refrain from laughing? That would be playing with words, or, rather, misusing them" (in Detlefsen 2008, pp. 112–113).

[31] A discussion of impossibility proofs in the Nineteenth Century can be found in Lützen (2009).

[32] Especially for the impossibility of solving the quintic equation and for the case of geometric constructions, a long discussion can be found in Vuillemin (1963, Chapters 2–4).

[33] Crippa (2014), especially the Introduction.

[34] Heeffer (2008, p. 149), with smaller modifications.

formulations, analysis proceeded by assuming the existence of a solution for the problem under investigation, without initially worrying about the very possibility of constructing or exhibiting such a solution by the methods available to the problem-solver. As a consequence of this way of reasoning, impossibilities may occur among the envisioned outcomes of analysis. This conclusion had already been stressed by Pappus, but was also stressed by his early modern readers too, as happened in Gregory's *VCHQ*, for instance.

Another example that perfectly illustrates this situation is the incommensurability between the side and the diagonal of a square. Here, we proceed by assuming the existence of the ratio between the side and the diagonal of a square in order to discover that such a ratio is, in the end, an impossible object. This example was universally known among early modern geometers, and it was certainly not by chance that it was also the model followed by Gregory's and Leibniz's proofs of the impossibility of squaring the central conic sections. In the case of the latter impossibility proofs, finding a contradiction required more mathematical sophistication and the use of subtle algebraic manipulations. However, both Gregory's and Leibniz's arguments followed the same approach provided by the method of analysis. Moreover, as we have seen, Wallis's analytical operations leading to impossibility results (which include the quadrature of the circle) behave according to the same structure. While the reference to impossibility can be found in ancient characterisations of analysis, the analysis of the "moderns," namely, early-modern algebra, differs from its traditional counterpart insofar as the former involved the possibility of introducing new symbols in order to denote impossible operations and results. The possibility of representing, through an adequate symbolism, impossibility results was unknown to classical geometrical reasoning, confined within the limits of diagrammatic representations in the plane or the space.

So, it's on these two levels, operational and methodological, that algebra paved the way for the study of impossibility results in the Seventeenth Century in an unprecedented way. James Gregory's words that a new realm of mathematics would be opened by the study of impossibility are prophetical when looking at the later history of Nineteenth and Twentieth Century mathematics. Yet, they are insightful even solely considered in their own context. The question of impossibility was a phenomenon that ancient geometers could describe, but only in very few cases were impossibility claims the subject matter of theorems and proofs (for instance, with the study of irrationality). On the other hand, especially in the second half of the Seventeenth Century, impossibility results became a legitimate subject of inquiry for mathematicians. This aspect of the scientific revolution of the Seventeenth Century has never been addressed thoroughly, and I hope my study has helped to shed some light upon it.

References

Alexander, Amir. 2014. *Infinitesimals: how a dangerous mathematical theory shaped the modern world.* New York: Farrar, Strauss and Giroux.

Andersen, Kirsti. 1984. Cavalieri's method of indivisibles. *Archive for History of Exact Sciences.* 31: 291–267.

Archimedes. 1881. *Archimedis Opera Omnia cum commentariis Eutociis.* 3 vols. Ed. J.L. Heiberg. Leipzig: Teubner.

Arthur, Richard. 2001. *The labyrinth of the continuum. Writings on the Continuum Problem, 1672-1686.* Yale: Yale University Press.

Arthur, Richard. 2013. Leibniz's syncategorematic infinitesimals. *Archive for History of Exact Sciences.* 67(5): 553–593.

Baron, Margaret. 1969. *The Origins of infinitesimal calculus.* New York: Dover.

Beeley Philip and Scriba, Christoph. 2008. Disputed Glory. John Wallis and some questions of precedence in seventeenth-century mathematics. In *Kosmos und Zahl. Beiträge zur Mathematik- und Astronomiegeschichte, zur Alexander von Humboldt und Leibniz.* 275–299. Ed. H. Hacht, R. Mikosch, I. Schwarz et. al, 275–299. Stuttgart: Franz Steiner Verlag.

Bernard, Alain. 2003. Sophistic Aspects of Pappus's *Collection. Archive for History of Exact Sciences.* 57: 93–150.

Blåsjö, Viktor. 2017. *Transcendental curves in the Leibnizian calculus.* Amsterdam: Elsevier.

Borwein, Jonathan and Borwein, Peter. 1998. *π and the AGM: A Study in Analytic Number Theory and Computational Complexity.* New York: Wiley-Interscience.

Bos, Henk. 1981. On the representation of curves in Descartes' Géométrie. *Archive for History of Exact Sciences.* 24: 295–338.

Bos, Henk. 1984. Arguments on Motivation in the Rise and Decline of a Mathematical Theory; the "Construction of Equations", 1637 - ca 1750. *Archive for History of Exact Sciences.* 30: 331–380.

Bos, Henk. 2001. *Redefining geometrical exactness.* Dordrecht Heidelberg New York London: Springer.

Breger, Herbert. 1986. Leibniz' Einführung des Transzendenten. *Studia Leibnitiana Sonderheft.* Ed. A. Heinekamp. 14: 119–132.

Brouncker, William. 1668. The squaring of the hyperbola by an infinite series of rational numbers, together with its demonstration by the Right Honourable the Lord Viscount Brouncker. *Philosophical Transactions of the Royal Society.* 3: 645–649.

D. Crippa, *The Impossibility of Squaring the Circle in the 17th Century*,
Frontiers in the History of Science, https://doi.org/10.1007/978-3-030-01638-8

Burn, Robert. 2001. Alphonse Antonio de Sarasa and Logarithms. *Historia Mathematica.* 28(1): 1–17.

Cajori, Florian. 1913. History of the exponential and logarithmic concepts. *The American Mathematical Monthly* 20(2): 35–47.

Child, J. M. 1920. *The early mathematical manuscripts of Leibniz*. Chicago: The Open Court Publishing Company.

Clagett, Marshall. 1964. *Archimedes in the Middle Ages. Volume 1.* Madison: The University of Wisconsin Press.

Cortese, João. 2016. When two points coincide, or are at an infinitely small distance: some aspects of the relation between the works of Leibniz, Pascal (and Desargues). In *Für unser Glück oder das Glück anderer: Vorträge des X. Internationalen Leibniz-Kongresses*, vol. 4. Ed. Wenchao Li, 165–178. Hildesheim: Olms.

Crippa, Davide. 2014. *Impossibility results: from geometry to analysis.* Université Paris Diderot: Phd Dissertation.

Crippa, Davide. 2017. Review of Mathematische Schriften, Reihe 7, sechster Band, 16731676, Arithmetische Kreisquadratur, Gottfried Wilhelm Leibniz. Akademie-Verlag, Berlin (2012), Siegmund Probst and Uwe Mayer, Eds. *Historia Mathematica.* 44, 1, 73–76.

Cuomo, Serafina. 2000. *Pappus of Alexandria and the mathematics of Late Antiquity*. Cambridge: Cambridge University Press.

Dehn, Max and Hellinger, Ernst. 1939. On James Gregory's "Vera quadratura". In *The James Gregory tercentenary memorial volume*. Ed. Herbert W. Turnbull, 468–478. Edinburgh: Royal Society of Edinburgh.

Dehn, Max and Hellinger Ernerst. 1943. Certain mathematical achievements of James Gregory. In *The American Mathematical Monthly.* 50(3): 149–163.

Debuiche, Valérie. 2013. L'expression Leibnizienne et ses modèles mathématiques. *Journal of the History of Philosophy* 51(3): 409–459.

Descartes, René. 1897–1913. *Oeuvres de Descartes*. Eds. Charles Adam and Paul Tannery. 12 vols. Paris: Cerf.

Descartes, René. 1952. *The Geometry of René Descartes*. Ed., transl. David E. Smith and Marcia L. Latham. La Salle: Open Court.

Descartes, René. 1659-1661. *Renati Descartes Geometria. Editio Secunda. Multis accessionibus exornata, et plus altera sua parte adaucta.* Ed. Frans Van Schooten. Amsterdam: Apud Ludovicum et Danielem Elzevirios.

Detlefsen, Michael. 2008. Interview with Michael Detlefsen. In *Philosophy of Mathematics. 5 Questions*. Eds. Vincent F. Hendricks and Hannes Leitgeb. Copenhagen: Automatic Press.

Dijksterhuis, E. Jan. 1939. James Gregory and Christiaan Huygens. In *The James Gregory Tercentenary Memorial Volume*. Ed. Herbert W. Turnbull, 478–486. Edinburgh: Bell and Sons.

Dijksterhuis, E. Jan. 1987. *Archimedes, with a new bibliographic essay.* ed. Wilbur R. Knorr. Princeton: Princeton University Press.

Dhombres Jean. 2014. A tentative interpretation of the epistemological significance of the encrypted message sent by Newton to Leibniz in October 1676. *Advances in Historical Studies* 3(1): 22–32.

Dutka, Jacques. 1982. Wallis's product, Brouncker's continued fraction, and Leibniz's series. *Archive for History of Exact Sciences* 26(2): 115–126.

Edward, Charles H. Jr. 1994. *The historical development of the calculus*. New York: Dover.

Fabri, Honoré. 1659. *Opusculum geometricum de linea sinuum et cycloide.* Roma: Typis HH. Francisci Corbelletti.

Feingold, Mordechai. 1996. Huygens and the Royal Society. *De Zeventiende Eeuw.* 12: 22–36.

Ferraro, Giovanni and Panza, Marco. 2003. Developing into series and returning from series: A note on the foundations of eighteenth-century analysis. *Historia Mathematica.* 30: 17–46.

Ferraro, Giovanni. 2008. *The rise and development of the theory of series up to the early 1820s*. Dordrecht Heidelberg New York London: Springer.

Gavagna, Veronica. 2014. *Radices Sophisticae, Racines Imaginaires*: The Origins of Complex Numbers in the Late Renaissance. In *The art of science. From perspective drawings to Quantum randomness*. Eds. Rossella Lupacchini, Annarita Angelini, 165–190. Dordrecht Heidelberg New York London: Springer.

Girard, Albert. 1629. *Invention nouvelle en l'algebre*. Amsterdam: Guillaume Iansson Blaeuw.

Giusti, Enrico. 1980. *Bonaventura Cavalieri and the Theory of Indivisibles*. Bologna: Edizioni Cremonese.

Gregory, James. 1667. *Vera circuli et hyperbolae quadratura, in sua propria specie inventa.* Padua: ex typographia Iacobi de Cadorinis.

Gregory, James. 1668. *Geometriae pars universalis*. Padua: Paolo Frambotti.

Gregory, James. 1668a. Mr. Gregories answer to the animadversions of Mr. Hugenius upon his book, De vera Circuli & Hyperbolae Quadratura; as they were publish'd in the Journal des Scavans of July 2. 1668. *Philosophical Transactions of the Royal Society.* 3, 37: 732–735.

Gregory James. 1668b. *Exercitationes geometricae*. London: William Godbid.

Gregory, James. 1668c. An Extract of a Letter of Mr. James Gregory to the Publisher, Containing Some Considerations of His, upon M. Hugens His Letter, Printed in Vindication of His Examen of the Book, Entitled Vera Circuli & Hyperbola Quadratura. *Philosophical Transactions of the Royal Society.* 3, 44: 882–886.

Grootendorst, A. W., and Van Maanen, Jan. 1982. Van Heuraet's letter (1659) on the rectification of curves. Text, translation (English, Dutch), commentary. *Niew Archief voor Wiskunde.* 30: 95–113.

Boas Hall, Marie. 2002. *Henry Oldenburg: shaping the Royal Society.* Oxford: Oxford University Press.

Heath, Thomas. 1896. *Apollonius of Perga treatise on conic sections (edited in modern notation).* Cambridge: Cambridge University Press.

Heath, Thomas. 1897. *The Works of Archimedes*. Cambridge: Cambridge University Press.

Heath, Thomas. 1921. *A history of Greek mathematics*. Oxford: Clarendon Press.

Heeffer, Albrecht. 2008. The Emergence of Symbolic Algebra as a Shift in Predominant Models. *Foundations of science*, 13(2): 149–161.

Hofmann, Joseph. 1942. Die Quellen der cusanischen Mathematik I: Ramon Lulls Kreisquadratur. *Sitzungsberichte der Heidelberger Akademie der Wissenschaften Philosophisch-historische Klasse.* 4: 1–38.

Hofmann, Joseph. 1974. *Leibniz in Paris (1672-1676). His growth to mathematical maturity.* Trans. A. Prag and D. T. Whiteside. Cambridge: Cambridge University Press.

De l'Hôpital, Guillaume François Antoine. 1707. *Traité analytique des sections coniques et de leur usage pour la résolution des équations dans les problèmes tant déterminés qu'indéterminés.* Paris: Montalant.

Horvath, Miklos. 1983. On the Leibnizian quadrature of the circle. *Annales universitatis scientiarum budapestiensis (Sectio computatorica).* 4: 75–83.

Huygens, Christiaan. 1668. Extrait d'une lettre de M. Huigens à l'auteur du Journal, touchant la réponse que M. Gregory a faite à l'examen du livre intitulé Vera Circuli et Hyperboles Quadratura, dont on a parlé dans le V. Journal de cette anée. *Journal des Sçavans.* 12 November 1668: 109–112.

Huygens, Christiaan. 1724. Christiani Hugenii Zulichemii, Dum viveret Zelemii Toparchae, Opera Varia. Volumen primum. Lugduni Batavorum: apud Janssonios vander Aa.

Huygens, Christiaan. 1888–1950. *Oeuvres complètes publiées par la Société hollandaise des sciences*, 22 vol. Ed. Bierens de Haan. The Hague: M. Nijhoff.

Israel, Giorgio. 1998. The analytic method in Descartes' *Géométrie*. In *Analysis and synthesis in mathematics*. Eds. Michael Otte, Marco Panza, 5–30. Boston studies in the philosophy of science. Dordrecht: Kluwer.

Jacob, Marie. 2005. *La quadrature du cercle : Un problème à la mesure des Lumières*. Paris: Fayard.

Jesseph, Douglas. 1999. *Squaring the Circle: the War between Hobbers and Wallis*. Chicago and London: University of Chicago Press.

Jones Arthur, Morris Sidney, Pearson Kenneth. 1991. *Abstract algebra and famous impossibilities*. New York: Springer.

Journal. 1668. Vera Circuli et Hyperboles Quadratura, in propria sua proportionis specie inventa et demonstrata a Jacobo Gregorio Scoto. In 4 Patavii (anonymous review). *Le Journal des Sçavans de l'an 1668*: 74.

Jullien, Vincent (editor). 2015. *Seventeenth-century indivisibles revisited*. New York, Dordrecht, London: Birkhäuser Basel.

Keyser, Paul T. 2007. Ammonius. In: *The Biographical Encyclopedia of Astronomers*. Eds. Hockey T. et al. New York: Springer.

Klein, Felix. *Famous problems of elementary geometry*, transl. W.W. Beman and E.D. Smith. Boston, London: Ginn and Company.

Knobloch, Eberhard. 1989. Leibniz et son manuscrit inédit sur la quadrature des sections coniques. In *The Leibniz Renaissance, International Workshop (Florence 2–5, 1986)*, ed. Ferdinando Abbri, 127–151. Firenze: L. Olschki.

Knobloch, Eberhardt. 2002. Leibniz's rigorous foundation of infinitesimal geometry by means of Riemannian sums. *Synthese.* 133(1): 59–73.

Knobloch, Eberhardt. 2008. Generality and Infinitely Small Quantities in Leibniz's Mathematics: The Case of his Quadrature of Conic Sections and Related Curves. In *Infinitesimal Differences: Controversies between Leibniz and his Contemporaries*. Eds. Ursula Goldenbaum, Douglas Jesseph, 171–183. Berlin-New York: Walter de Gruyter.

Knorr, Wilbur Richard. 1983. Construction as existence proof in Ancient Geometry. *Ancient Philosophy* 3: 125–148.

Knorr, Wilbur Richard. 1986. *The Ancient Tradition of Geometric Problem Solving*. New York, Dordrecht, London: Birkhäuser Basel.

Lehay, Andrew. 2016. William Neile's Contribution to Calculus. *The College mathematics journal* 46 (1): 42–49.

Leibniz, Gottfried Wilhelm. 1849–1863. *Mathematische Schriften. 7 vols*. Ed. Carl I. Gerhardt. Berlin: Schmidt (Reprinted 1971-, Hildesheim: Olms).

Leibniz, Gottfried Wilhelm. 1923-. *Sämtliche Schriften und Briefe*. Berlin, Göttingen: Berlin-Brandenburgische Akademie der Wissenschaften/Akademie der Wissenschaften zu Göttingen.

Leibniz, Gottfried Wilhelm. 1989. *Naissance du calcul différentiel*. Ed. Marc Parmentier. Paris: Vrin.

Leibniz, Gottfried Wilhelm. 1984. *Philosophical essays*. Eds. Roger Ariew, Daniel Garber. Indianapolis & Cambridge: Hackett Publishing Company.

Leibniz Gottfried Wilhelm. 2004. *Quadrature arithmétique du cercle, de l'ellipse et de l'hyperbole et la trigonométrie sans tables qui en est le corollaire*. Ed. Marc Parmentier. Paris: Vrin.

Leibniz, Gottfried Wilhelm. 2011. *Gottfried Wilhelm Leibniz: Die mathematischen Zeitschriftenartikel*. Eds. Hans-Jürgen Hess, and Malte-Ludof Babin. Hildesheim, Zürich, New York: Georg Olms Verlag.

Leibniz, G. W. 2015. *Obras filosóficas y cientificas. Escritos matemáticos,* 7A, ed. M. S. de Mora Charles. Granada: Editorial Comares.

Leibniz, G. W. 2016. De quadratura arithmetica circuli ellipseos et hyperbolae cujus corollarium est trigonometria sine tabulis, ed. E. Knobloch (tr. O Hamborg). Berlin and Heidelberg: Springer Spektrum.

Leibniz, Gottfried Wilhelm. 2018. *Ecrits sur la mathesis universalis*. Ed. David Rabouin. Paris: Vrin.

Levey, Samuel. 2008. Archimedes, Infinitesimals and the Law of Continuity: On Leibniz's Fictionalism. In *Infinitesimal differences: Controversies between Leibniz and his Contemporaries*. Eds. Ursula Goldenbaum, Douglas Jesseph. 107–133. Berlin-New York: Walter de Gruyter.

Lützen, Jesper. 2009. Why was Wantzel overlooked for a century? The changing importance of an impossibility result. *Historia Mathematica*. 36: 374–394.

Lützen, Jesper. 2010. The algebra of geometric impossibilities: Descartes and Montucla on the impossibility of the duplication of the cube and the trisection of the angle. *Centaurus*. 52: 4–37.

Lützen, Jesper. 2014. 17th century arguments for the impossibility of the indefinite and the definite quadrature of the circle. *Revue d'histoire des mathématiques*. 20: 211–251.

Van Maanen, Jan. 1984. Hendrick van Heuraet (1634-1660?): His Life and Mathematical Work. *Centaurus*. 27, 3–4: 218–279.

Mahnke, Dietrich. 1925. Neue Einblicke in die Entdeckungsgeschichte der höheren Analyse. *Abhandlungen der Preussische Akademie der Wissenschaften*. Phys-math Klass, 1: 1–71.

Mancosu, Paolo. 1999. *Philosophy of Mathematics and Mathematical Practice in XVII Century*. Oxford: Oxford University Press.

Mancosu Paolo. 2007. Descartes and mathematics. In *A companion to Descartes*. Ed. Janet Broughton, John Carriero, 103–123. Oxford: Blackwell Publishing.

Malet, Antoni. 1989a. *Studies on James Gregorie*. Phd Dissertation: Princeton University.

Malet, Antoni. 1996. *From Indivisibles to Infinitesimals: Studies on Seventeenth-Century Mathematizations of Infinitely Small Quantities*. Barcelona: Universitat Autonoma de Barcelona.

Malet, Antoni and Panza, Marco. 2015. Wallis on indivisibles. In *Seventeenth-century indivisibles revisited*. Ed. Vincent Jullien. 307–347. New York, Dordrecht, London: Birkhäuser Basel.

Massa Esteve, Maria Rosa. 2016. Mengoli's mathematical ideas in Leibniz's excerpts. *BSHM Bulletin: Journal of the British Society for the History of Mathematics*. 32, 1: 1–21.

Mengoli, Pietro. 1650. *Novae quadraturae arithmeticae*. Bologna: Montij.

Mercator, Nicolaus. 1668. *Logarithmotechnia, sive methodus construendi logarithmos nova, accurata et facilis*. London: William Godbid.

Molland, George A. 1991. Implicit versus explicit geometrical methodologies: the case of constructions. In *Mathématiques et philosophie de l'antiquité à l'âge classique: hommage à Jules Vuillemin*. Ed. R. Rashed. 181–196. Paris: editions du CNRS.

Montucla, Jean-Etienne. 1831. *Histoire des recherches sur la quadrature du cercle, avec une addition concernant les problèmes de la duplication du cube et de la trisection de l'angle*. Ed. François S. Lacroix. Paris: Bachelier père et fils.

Panza, Marco. 2005. *Newton et les origines de l'analyse*. Paris: Blanchard.

Panza, Marco. 2011. Rethinking geometrical exactness. *Historia Mathematica*. 38: 42–95.

von Pape, Bodo. 2017. Diedrich Uhlhorn (1764-1837) und die Grössen Probleme der Antike. *Mitteilungen der Hamburger Mathematischen Gesellschaft*. 27: 29–59.

Pappus. 1876. *Pappi Alexandrini collectionis quae supersunt*, 3 vols. Ed. Friedrich Otto Hultsch. Berlin: Weidman.

Pappus. 1986. *Book 7 of the Collection, edited by A. Jones*. ed., transl. Alexander Jones. Dordrecht Heidelberg New York London: Springer.

Pappus. 2010. *Pappus of Alexandria: Book 4 of the Collection*. ed., transl. Heike Sefrin Weis. Dordrecht Heidelberg New York London: Springer.

Pasini, Enrico. 1993. *Il reale e l'immaginario. La fondazione del calcolo infinitesimale nel pensiero di Leibniz*. Torino: Edizioni Sonda.

Probst, Siegmund. 2006. Zur Datierung von Leibniz' Entdeckung der Kreisreihe. In *Einheit in der Vielheit. VIII. Internationaler Leibniz-Kongress*. Eds. Herbert Breger, Jürgen Herst and Sven Erdner. 813–817. Hannover: Gottfried-Wilhelm-Leibniz-Gesellschaft.

Probst, Siegmund. 2008. Indivisibles and Infinitesimals in Early Mathematical Texts of Leibniz. In *Infinitesimal differences: Controversies between Leibniz and his Contemporaries*. Eds. Ursula Goldenbaum, Douglas Jesseph. 95–106. Berlin-New York: Walter de Gruyter.

Probst, Siegmund. 2015. Leibniz as Reader and Second Inventor: The Cases of Barrow and Mengoli. In *G. W. Leibniz, interrelations between mathematics and philosophy*. Eds. Norma B. Goethe, Philip Beeley, David Rabouin. 111–134. Dordrecht Heidelberg New York London: Springer.

Probst, Siegmund. 2016. Leibniz und Roberval. In *Für unser Glück oder das Glück anderer. Vorträge des X. Internationalen Leibniz-Kongresses*. Eds. Wenchao Li et al. Band IV. 183–189. Hildesheim: Olms.

Rabouin, David. 2015. Leibniz's rigorous foundation of the method of indivisibles. In *Seventeenth-century indivisibles revisited*. Ed. Vincent Jullien. 347–365. New York, Dordrecht, London: Birkhäuser Basel.

Rashed, Roshdi. 2013. *Ibn al-Haytham and analytic mathematics*. Trans. Daniel O'Donoghue. Ed. Geoffrey Nash. London, New York: Routledge.

Rosso, Riccardo. 2014. Appunti di storia dei logaritmi V: logaritmi e serie. *L'insegnamento della Matematica e delle Scienze Integrate*. 37B: 311–342.

Sasaki, Chihara. 2003. *Descartes's mathematical thought*. Dordrecht Heidelberg New York London: Springer.

Scriba, Christoph. 1957. *James Gregorys frühe Schriften zur Infinitesimalrechnung*. Giessen: Selbstverlag des Mathematischen Seminars.

Scriba, Christoph. 1983. Gregory's converging double sequence. A New look at the controversy between Huygens and Gregory over the "analytical" quadrature of the circle. *Historia Mathematica*. 10: 274–285.

Smeur Alphons Johannes Emile Marie. 1970. On the value equivalent to π in ancient mathematical texts. A new interpretation. *Archive for History of Exact Sciences*. 6(4): 249–270.

Stedall, Jacqueline. 2002. *A Discourse Concerning Algebra: English Algebra to 1685*. Oxford: Oxford University Press.

Stedall, Jacqueline. 2008. *Mathematics emerging. A sourcebook: 1540–1900*. Oxford: Oxford University Press.

Stevin, Simon. 1958. *The principal works of Simon Stevin*. Ed. D. J. Struik. Amsterdam: Swets & Zeitlinger.

Szabó Árpád. 1978. *The beginnings of Greek mathematics*. Dordrecht, Heidelberg, New York, London: Springer.

Transactions. 1668. An account of some Books (anonymous review). *Philosophical Transactions of the Royal Society*. 3: 685–688.

Tropfke Johannes. 1902. *Geschichte der Elementar-Mathematik in systematischer Darstellung. Zweiter Band*. Leipzig: Verlag von Veit & Comp.

Turnbull, Herbert W. 1939. *The James Gregory Tercentenary Memorial Volume*. Edinburgh: Bell and Sons.

Viète François. 1593. *Variorum de rebus mathematicis responsorum Liber VIII*. Tour: Mettayer.

Viète François. 1646. *Francisci Vietae opera mathematica*, in unum Volumen congesta, ac recognita, opera atque studio Francisci a Schooten Leydensis, Matheseos professoris Lugduni Batavorum. Leyden: Ex officina Johannis Elsevirii.

Viète, François. 1983. *The Analytic Art: Nine Studies in Algebra, Geometry and Trigonometry from the Opus restitutae Mathematicae Analyseos, seu Algebra Nova*. Ed. R. Witmer. New York: Dover.

Euclid. 1990. *Les Eléments: traduction et notes de Bernard Vitrac. Volume III*. Ed. Bernard Vitrac. Paris: Presses Universitaires de France (PUF).

Vuillemin Jules. 1963. *La philosophie de l'algèbre: t. I: Recherches sur quelques concepts et méthodes de l'algèbre moderne*. Paris: Epimethée.

Yoder, Joella. 1988. *Unrolling time. Christiaan Huygens and the mathematization of nature*. Cambridge: Cambridge University Press.

Youkshevitch A. P. 1976. The Concept of Function up to the Middle of 19th Century. *Archive for History of Exact Sciences*. 16(1): 37–85.

Wallis, John. 1659. *Tractatus duo, prio de cycloide et corporibus inde genitis. Posterior, epistolaris, in qua agitur de cissoide et corporibus inde genitis*, Oxford: typis Academicis Lichfieldianis.

Wallis, John. 1668. An account of two books. *Philosophical Transactions of the Royal Society*, 3: 640–644.

Wallis, John. 1685. *A treatise of algebra both historical and practical shewing the original, progress, and advancement thereof, from time to time; and by what steps it hath attained to the height at which now it is*. London.

Wallis, John. 2004. *The arithmetic of infinitesimals*. Ed., transl. Jacqueline Stedall. Dordrecht Heidelberg New York London: Springer.

Wallis, John. 2003. *The Correspondence of John Wallis (1616-1703). Volume I*. Eds. Philip Beeley, Christoph Scriba. Oxford: Oxford University Press.

Wallis, John. 2012. *The Correspondence of John Wallis (1616-1703). Volume III. October 1668-1671*. Eds. Philip Beeley, Cristoph Scriba. Oxford: Oxford University Press.

Wallis, John. 2014. *The Correspondence of John Wallis (1616-1703). Volume IV. 1672-April 1675*. Eds. Philip Beeley, Cristoph Scriba. Oxford: Oxford University Press.

Whiteside, Derek, Thomas. 1961. Patterns of mathematical thought in later 17th century mathematics. *Archive for History of Exact Sciences*. 1: 179–388.

Zeuthen, H. Georg. Die geometrische Konstruktion als "Existenzbeweis" in der antiken Geometrie. *Mathematische Annalen*. 47: 222–228.

Zeuthen, H. Georg. 1903. *Geschichte der Mathematik im XVI. and XVII. Jahrhundert*. transl. R. Meyer. Leipzig: Teubner.

Index

D. Crippa, *The Impossibility of Squaring the Circle in the 17th Century*,
Frontiers in the History of Science, https://doi.org/10.1007/978-3-030-01638-8

Printed in the United States
By Bookmasters